Lecture Notes in Mathematics

Edited by A. Dold and B. Eckmann

Subseries: Fondazione C.I.M.E., Firenze
Adviser: Roberto Conti

1048

Kinetic Theories and the Boltzmann Equation

Lectures given at the 1st 1981 Session of the
Centro Internazionale Matematico Estivo (C.I.M.E.)
Held at Montecatini, Italy, June 10 – 18, 1981

T0226100

Edited by C. Cercignani

Springer-Verlag
Berlin Heidelberg New York Tokyo 1984

Editor

Carlo Cercignani
Dipartimento di Matematica
Politecnico di Milano, Milano, Italy

AMS Subject Classifications (1980): 76P05, 82A40, 82A45, 82A70

ISBN 3-540-12899-9 Springer-Verlag Berlin Heidelberg New York Tokyo
ISBN 0-387-12899-9 Springer-Verlag New York Heidelberg Berlin Tokyo

Library of Congress Cataloging in Publication Data. Main entry under title: Kinetic theories and the Boltzmann equation. (Lecture notes in mathematics; 1048) 1. Transport theory–Congresses. 2. Gases, Kinetic theory of–Congresses. 3. Evolution equations–Congresses. I. Cercignani, Carlo. II. Centro internazionale matematico estivo. III. Series.
QC175.2.A1K56 1984 530.1'36 84-3118
ISBN 0-387-12899-9 (U.S.)

Printing and binding: Beltz Offsetdruck, Hemsbach/Bergstr.
2146/3140-543210

PREFACE

The book contains the text of three of the four series of lectures plus a few seminars presented at the first session of the Summer School organized at C.I.M.E. We regret very much that we were unable to obtain the text of the lectures on the singular limits of the Boltzmann Equation by Professor Harold Grad, in spite of the fact that we waited for them more than two years.

The unifying theme of the School was the study of evolution equations whose unknown is a distribution function describing the probabilistic behaviour of the underlying particle dynamics.

The lecture notes presented in this volume deal with the time dependent linear transport equation (by Professor J.J. Hejtmanek of Vienna University), the existence theorems for the nonlinear Boltzmann equation (by Professor P.F. Zweifel of Virginia Polytechnic Institute and State University), the Boltzmann-Vlasov equation for ionized gases (by Professor H. Neunzert of the University of Kaiserslautern).

The text of the seminars deals with half space problems for kinetic models, the Boltzmann equation for molecular forces of infinite range, a survey of recent results of the Cauchy problem for the Boltzmann equation, the Boltzmann hierarchy.

Except for the forced omission mentioned above, we feel that the volume gives a coherent picture of this field of applied mathematics.

Carlo Cercignani

C.I.M.E. Session on Kinetic Theories and Boltzmann Equation

List of participants

G. Albano, Istituto di Fisica, Univ. di Palermo, Palermo

L. Arlotti, Istituto di Matematica, via della Montagnola 30, 60100 Ancona

M. Arthur, Abteilung f. Math. Phys., Univ. Ulm, 7900 Ulm.

F. Bampi, Istituto di Matematica, via L.B. Alberti 4, 16132 Genova

I. Barbieri, Istituto Matematico, Piazza di Porta S. Donato 5, 40128 Bologna

M.F. Barnsley, School of Math., Georgia Inst. of Technology, Atlanta, GA 30332

L. Bassi, Istituto di Matematica, via della Montagnola 30, 60100 Ancona

M. Benati, Istituto di Matematica, via L.B. Alberti 4, 16132 Genova

G. Boillat, 9, rue d'Anvers, 25000 Besançon

G. Borgioli, Istituto di Matematica Applicata "G.Sansone", v.le Morgagni 44, 50134 Firenze .

G. Busoni, Istituto di Matematica, via della Montagnola 30, 60100 Ancona

V. Capasso, Istituto di Matematica, Palazzo Ateneo, 70121 Bari

L. Caprioli, Istituto di Matematica Applicata, via Vallescura 2, 40136 Bologna

G.L. Caraffini, Istituto di Matematica, via dell'Università 12, 43100 Parma

G. Caviglia, Istituto di Matematica, via L.B. Alberti 4, 16132 Genova

C. Cercignani, Istituto di Matematica del Politecnico, P.za L. da Vinci 32, 20133 Milano

A. De Vito, Istituto di Matematica, via della Montagnola 30, 60100 Ancona

Vl. Ďurikovič, Dept. of Math. Analysis, Komensky University, Mat. pavilon, Mlynska dolina, 81631 Bratislava

T. Elmroth, Matematiska Institutiones, Chalmers TH, S 412 96 Göteborg

G. Ferrarese, Istituto Matematico, Univ. di Roma, Città Universitaria, 00100 Roma

F. Franchi, Istituto Matematico, P.za di Porta S. Donato 5, 40127 Bologna

G. Frosali, Istituto Matematico, V.le Morgagni 67/A, 50134 Firenze

H. Grad, Courant Institute, NYU, 251 Mercer Street, New York, NY 10012

A. Greco, Istituto di Matematica, Facoltà di Ingegneria, V.le delle Scienze, Palermo

J. Hejtmanek, Institut f. Math., Univ. Wien, Strudlhofgasse 4, A 1090 Wien

L. Lachowicz, Warsaw Univ., Dept. of Math. and Mech., PkiN IX p., 00901 Warszawa

M. Lampis, Istituto di Matematica del Politecnico, P.za L. da Vinci 32, 20133 Milano

A. Majorana, Seminario Matematico, V.le A. Doria 6, 95125 Catania

A. Morro, Istituto di Matematica, via L.B. Alberti 4, 16132 Genova

R. Nardini, Istituto Matematico, P.za di Porta S. Donato 5, 40127 Bologna

H. Neunzert, Univ. Kaiserslautern, Fachbereich Math., Pfaffenbergstrasse, 6750 Kaiserslautern

A. Pałczewski, Warsaw Univ., Dept. of Math. and Mech., PkiN IX p., 00901 Warszawa

P.S. Pantano, Dipartimento di Matematica, Università della Calabria, Arcavacata di Rende, Cosenza

S.L. Paveri Fontana, Ist. Univ. Navale, via Amm. F. Acton 38, 80133 Napoli

R. Pettersson, Matematiska Institutionen, Chalmers TH, S 41296 Göteborg

T. Płatkowski, Warsaw Univ., Dept. of Math. and Mech., PkiN IX p., 00901 Warszawa

Vl. Rosa, Dept. of Math. Analysis, Komensky University, Mat. pavilon, Mlynska dolina, 81631 Bratislava

T. Ruggeri, Istituto di Matematica Applicata, via Vallescura 2, 40136 Bologna

F.J. Schwarz, Fachbereich Math., Universität, 6750 Kaiserslautern

H. Spohn, Theoretische Physik, Theresienstr. 37, 8 München 2

C. Tebaldi, Laboratorio di Fisica Nucleare, Montecuccolino, via dei Colli 16, 40136 Bologna

S. Totaro, Istituto Matematico, viale Morgagni 67/A, 50134 Firenze

L. Triolo, Istituto di Matematica Applicata, via A. Scarpa 10, 00161 Roma

G. Turchetti, Istituto di Fisica, via Irnerio 46, 40100 Bologna

M. Wolska, Instituto of Mathematics, ul. St. Banacha 22, 90-238 Łodz

T. Ytrehus, Dept. of Aeronautical Engng., Univ. of Illinois, 104 S.Mathews Ave., Urbana, Ill. 61801

C. Zanin, Istituto di Matematica, via Mantica 1, 33100 Udine

C. Zordan, Istituto di Matematica, via L.B. Alberti 4, 16132 Genova

P.F. Zweifel, Virginia Polytechnic Inst., Lab. for Transport Theory and Math. Phys., Blacksburg, VA 24061

TABLE OF CONTENTS

TIME-DEPENDENT LINEAR TRANSPORT THEORY

J. Hejtmanek

§ 1 TIME DEPENDENT LINEAR TRANSPORT EQUATIONS

Let us assume that we are dealing with a system of particles of two different species: host particles and test particles. At all times the host particles maintain a known equilibrium distribution which is not affected by the test particles. The density of the test particles is sufficiently low relative to that of the host particles that interactions among the test particles can be ignored. The only interactive process contributing to the rate of change of the test particle distribution is that between a test particle and a host particle - which is a linear process, as far as the density of the test particles is concerned. The basic quantity of interest is the test particle distribution function f, which contains information on the spatial distribution of the particles in configuration space and on the velocity distribution of the particles at each point of configuration space. The direct product of the configuration space and the velocity space is referred to as the phase space, so f is defined on a subset in phase space. Being a distribution function, f is real-valued and nonnegative. We assume that the region in configuration space Ω, where the transport processes take place, is an open convex subset of $\mathbb{R}^3$ and the velocity domain of the particles S is a ball or a spherical shell in $\mathbb{R}^3$.

We assume that the test particles move about freely (i.e., rectilinearly and with constant velocity) until they interact with a host particle of the transport system; in the course of an interaction a test particle may disappear entirely (absorption), it may change its velocity (scattering), or it may trigger a multiplication process, as a result of which one or more new test particles appear. The relevant space and time scales are such that interactions can be viewed as localized and instantaneous events. The expected number of the test particles in a volume element dx centered at a point $x \in \Omega$ whose velocities lie in a velocity element $d\xi$ centered at the velocity $\xi \in S$ at time t is represented by $f(x,\xi,t)dx\, d\xi$. The time dependent linear transport equation is a balance equation for f over the element $dx\, d\xi$ about (x,ξ),

$$(1.1) \quad \frac{\partial f}{\partial t} = -\frac{\partial}{\partial x}\cdot \xi\, f(x,\xi,t) - h(x,\xi)\cdot f(x,\xi,t) + \\ + \int_S k(x,\xi \leftarrow \xi')\, f(x,\xi',t)d\xi', \quad (x,\xi)\in \Omega \times S, \ t>0.$$

The first term on the right is the spatial divergence of the particle flux, which represents the effect of free streaming; the second term represents the loss due to interactions at x, $h(x,\xi)d\xi$ being the collision frequency for test particles with the velocity in the range $d\xi$ about ξ at the point x; the third term represents the gain due to interactions at x, $k(x,\xi \leftarrow \xi')d\xi$ being the expected number of test particles emerging with a velocity in the range $d\xi$ about ξ after an interaction of a test particle with the velocity ξ' with a host particle of the transport system at x. With Eq. (1.1) are prescribed an initial condition,

$$(1.2) \qquad f(x,\xi,0) = f_0(x,\xi), \quad (x,\xi)\in \Omega \times S,$$

and a boundary condition on $\partial\Omega$. Eq. (1.2), together with the initial condition (1.2) and the boundary condition, represents a linear Cauchy problem. We shall describe in more detail two model equations, the neutron transport equation, as it is applied to neutron experiments and in reactor physics, and the photon transport equation, as it is applied in Computerized Tomography.

The fundamental problem of reactor physics is the determination of the time behavior of a nuclear reactor; of special interest is the asymptotic behavior. The reactor which covers the bounded open convex subset Ω of R^3 is a highly heterogeneous composite structure of many different materials. The neutrons are the test particles. They move about freely until they interact with a nucleus of the reactor material. The nuclei are the host particles. It depends on the model, which subset of $\mathbb{R}^3$ is chosen for the velocity space. Thus, S can be S^2 in the case of the one group model, or S can be a ball or the whole space $\mathbb{R}^3$. The reaction rate function h can be computed from the macroscopic cross section functions.

$$(1.3) \qquad h(x,\xi) = |\xi| \, (\Sigma_a(x,\xi) + \Sigma_s(x,\xi) + \Sigma_f(x,\xi)), \quad (x,\xi) \in \Omega \times S,$$

when Σ_a, Σ_s and Σ_f are the macroscopic cross sections for absorption, scattering and fission processes. We might add that, for many reactor materials, the functions Σ_a, Σ_s and Σ_f vary rapidly with the neutron velocity ξ: they may display resonances, et cetera. Similarly the reaction rate function k can be computed,

$$(1.4) \qquad k(x,\xi \leftarrow \xi') = |\xi'| \, (\Sigma_s(x,\xi') \cdot \sigma_s(x,\xi \leftarrow \xi') +$$
$$+ \Sigma_f(x,\xi') \cdot \nu(x) \cdot \sigma_f(x,\xi \leftarrow \xi')), \quad (x,\xi) \in \Omega \times S,$$

where σ_s describes the velocity distribution of scattered neutrons and σ_f the corresponding velocity distribution of fission neutrons; ν is the mean number of neutrons produced in a fission process, and assumed to be independent of the velocity. Also the function k varies rapidly with the neutron velocity.

The boundary condition for the reactor problem expresses the fact that no neutrons enter the reactor from outside ("zero incoming flux"). It may be formulated as

$$(1.5) \qquad \xi f(x,\xi,t) = 0, \quad x \in \partial\Omega, \; \xi \in S_x, \; t > 0,$$

where $S_x := \{\xi \in S : x + t\xi \in \Omega \text{ for some } t > 0\}$, $x \in \partial\Omega$. We call the Cauchy problem (1.1), together with the initial condition (1.2) and the boundary condition of zero incoming flux (1.5) the reactor problem.

In general, the region $\Omega \times S$ in phase space is a sixdimensional manifold which, in case of the one group model (also called the monoenergetic model), reduces to the fivedimensional manifold $\Omega \times S^2$. If there are additional symmetry assumptions about the phase space, then the number of dimension can be reduced further. We mention the simplest model, the model of a homogeneous and isotropic slab reactor for monoenergetic neutrons and with azimuthal symmetry. In this case, the region $\Omega \times S$ reduces to a rectangle $(-a,a)\times[-1,1]$ in $\mathbb{R}^2$, and the linear transport equation has the form:

$$(1.6) \quad \frac{\partial f}{\partial t} = -\frac{\partial}{\partial x}\cdot \mu f(x,\mu,t) - h.f(x,\mu,t) + \\ + \frac{c}{2}\int_{-1}^{1} f(x,\mu',t)d\mu', \quad (x,\mu)\in(-a,a)\times[-1,1], \quad t>0,$$

with the initial condition

$$(1.7) \qquad f(x,\mu,0) = f_0(x,\mu), \quad (x,\mu)\in(-a,a)\times[-1,1],$$

and the boundary condition of zero incoming flux,

$$(1.8) \qquad \mu.f(-a,\mu>0,t) = 0 \text{ and } \mu.f(a,\mu<0,t) = 0, \quad t>0.$$

A scattering experiment for neutrons or other particles involves a piece of material, whose physical properties are the object of invertigation, a particle source, and an array of detectors placed at various positions around the target. In this case, we identify Ω with the entice space $\mathbb{R}^3$. We assume that the target occupies a region D which is a convex and compact subset of Ω. Inside D the particles collide with the host particles of the target. In these collision processes they can be absorbed or scattered, or they can produce new particles by fission. In a scattering process they change their velocity and direction; in general a particle will undergo more than one scattering process inside the region D. For this reason we refer to such an experiment as a multiple scattering experiment. Outside D the particles move freely, i.e. along straight lines with constant velocity. Thus, the region of interest in phase space is $\Omega \times S$, where $\Omega = R^3$, and the two reaction rate functions h and k which describe the transport process vanish outside D. We call the Cauchy problem (1.1), together with the initial condition (1.2) the multiple scattering problem. The choice of the function space in which the solution is sought reflects the boundary condition at infinity.

If we remove the target, which means that the two functions h and k are identically zero, the linear transport equation describes the motion of a cloud of particles in free space or vacuum,

$$\frac{\partial f}{\partial t} = -\frac{\partial}{\partial x}\cdot \xi f(x,\xi,t),\ (x,\xi)\in \Omega \times S,\ t>0. \tag{1.9}$$

Mathematical scattering theory is a comparison of the free Cauchy problem (1.9) and the interacting Cauchy problem (1.1). We notice that the term scattering is used here in two different meanings: a neutron is scattered in a collision process, which means that it undergoes a change in velocity and/or direction; in this sense we also use the term multiple scattering problem, namely, when neutrons may undergo more than one scattering process. The second meaning occours in the term scattering theory, or mathematical scattering theory, where it is used to describe the comparison between the free and the interacting dynamics. Scattering theory in this sense is part of the perturbation theory of linear operators - see, for example, Reed and Simon [1979].

In Computerized Tomography, a linear beam of photons from an X-ray source emerges from a collimator. The photons are absorbed during their flight along the straight line, and finally enter the X-ray detector via a second collimator. The photons are the test particles of our transport system, and the electrons of the human tissues are the host particles. The phase space of our problem is the manifold $R^2 \times S^1$, because we are interested to calculate the electron density function along a twodimensional cut through the human body, either through the brain or the abdomen. In our model, we assume that all photons have the same energy and travel with the velocity of light, $c = 1$. In this case, the linear transport equation has the following form:

$$(1.10) \quad \frac{\partial f}{\partial t} = -\frac{\partial}{\partial x}\cdot \xi \; f(x,\xi,t) - h(x).f(x,\xi,t), \; (x,\xi) \in \mathbb{R}^2 \times S^1, \; t > 0,$$

with the initial condition

$$(1.11) \qquad f(x,\xi,0) = f_0(x,\xi), \; (x,\xi) \in \mathbb{R}^2 \times S^1.$$

We notice that, in our model, the reaction rate function h depends only on x, i.e. the transport system is isotropic, but not homogeneous. We assume that h has support in a convex and compact subset of $\mathbb{R}^2$. We see that the Cauchy problem (1.10), together with the initial condition (1.11) is a special multiple scattering problem.

One quantity of interested in all linear transport systems is the total number of particles in the region of interest $\Omega \times S$ of the phase space:

$$(1.12) \qquad \int_\Omega \int_S f(x,\xi,t)dx\, d\xi, \; t \geq 0.$$

We notice that this integral is an L^1-norm, if the function $f(.,.,t)$ is realvalued and nonnegative. The choice of the space $L^1(\Omega \times S)$ for the functional formulation of our Cauchy problem is a natural one, as the L^1-norm of a nonnegative f gives the total number of neutrons in the reactor or multiple scattering problem and the total number of photons in the case of Computerized Tomography. We notice that $L^1(\Omega \times S)$ has a (real or complex) Banach space structure and a lattice structure. The positive cone $L_+^1(\Omega \times S)$ consists of all function f, which are realvalued and nonnegative almost everywhere in $\Omega \times S$. A necessary condition for the Cauchy problem is that the solution remains realvalued and nonnegative, if we start from a realvalued and nonnegative initial distribution f_0.

We have to distinguish between a linear and a linearized transport problem. Let us assume that there exists an equilibrium distribution function f_0, which solves a stationary nonlinear transport equation. If we can linearize this equation in a neighborhood of f_0, we arrive at a linear transport equation similar to Eq. (1.1). An equation of this kind is called linearized transport equation. Because the solution of this Cauchy problem is a perturbation of the equilibrium particle distribution function rather than a particle distribution function, it does not necessarily follow that the solution remains realvalued and nonnegative, if we start from a realvalued and nonnegative initial function.

In this series of lectures, we are concerned with the functional formulation of the reactor problem and the multiple scattering problem in the Banach lattice $L^1(\Omega \times S)$. We shall prove existence and uniqueness of the solution of the corresponding abstract Cauchy problem, which is given as a strongly continuous semigroup, we shall investigate the asymptotic behavior of the semigroup and the time dependent scattering theory of the multiple scattering problem.

§ 2 MATHEMATICAL PREREQUISITES

Let X be a Banach space, e.g. $X = L^1(\Omega \times S)$, and let $W = [W(t) : t \geq 0]$ be a strongly continuous semigroup of bounded operators in X. The type of the semigroup W is defined

$$(2.1) \qquad \omega_0(W) := \inf\{t^{-1} \ln\|W(t)\| : t > 0\} = \lim_{t\to\infty} t^{-1} \ln\|W(t)\|.$$

The generator $-T$ of the semigroup W is defined by

$$(2.2) \qquad \operatorname{dom} T := \{f \in X : \lim_{\delta\to 0} (W(\delta)f - f)/\delta \text{ exists}\} \text{ and}$$

$$-Tf := \lim_{\delta\to 0} (W(\delta)f - f)/\delta, \quad f \in \operatorname{dom} T.$$

We state two theorems which, together, are known as the Hille-Yosida theorem.

THEOREM 2.1. If $[W(t) : t \geq 0]$ is a strongly continuous semigroup whose type is $\omega_0(W)$, then: (i) the generator $-T$ is a closed linear operator with dense domain in X, (ii) for each $\omega > \omega_0(W)$ there exists an M such that the half-plane $\{\lambda \in \mathbb{C} : \operatorname{Re} \lambda > \omega\}$ is contained in the resolvent set of $-T$, and

$$(2.3) \qquad \|(\lambda I + T)^{-n}\| \leq M(\operatorname{Re} \lambda - \omega)^{-n}, \quad n = 1,2,3,\ldots$$

for each λ in the half-plane.

PROOF. See Dunford and Schwartz [1958, Section VIII. 1.13] or Kato [1966, Section IX.1] or Pazy [1974]. □

The constant M in (2.3) is independent of n; however, M may depend on ω. In general, (2.3) ceases to be valid at $\omega = \omega_0(W)$.

The resolvent of the generator of W can be calculated by means of the formula

$$(2.4) \qquad (\lambda I + T) = \int_0^\infty \exp(-\lambda t)W(t)dt, \ \mathrm{Re}\ \lambda > \omega_0(W).$$

The integral is the limit in the uniform topology of the integral

$$(2.5) \qquad \int_0^\tau \exp(-\lambda t)W(t)dt$$

as $\tau \to \infty$; the latter integral, in term, is defined as the strong limit of Riemann sums.

THEOREM 2.2. If T is a linear operator that satisfies the following conditions: (i) T is closed and dom T is dense in X, and (ii) there exists a real number ω and a positive number M such that (ω, ∞) is contained in the resolvent set of $-T$, and

$$(2.6) \qquad \|(\lambda I + T)^{-n}\| \leq M(\lambda - \omega)^{-n}, \ n = 1,2,3,\ldots,$$

for each $\lambda > \omega$, then there exists a strongly continuous semigroup W, whose generator is $-T$.

A linear operator T that satisfies the conditions (i) and (ii) of Theorem 2.2 will be called a Hille-Yosida operator.

Each operator W(t) of the semigroup W can be obtained from the generator $-T$ as the strong limit of a sequence of bounded operators

$$(2.7) \qquad W(t) = \lim_{n\to\infty} (I + \tfrac{t}{n}T)^{-n}.$$

One has the estimate $\|W(t)\| \leq M \exp(\omega t)$, where the constants M and ω are the same as in Theorem 2.2.

The semigroup W plays an important role in the solution of initial value problems for the abstract differential equation $u'(t) + Tu(t) = 0$ in the Banach space X. A vector-valued function $u: [0, \infty) \to X$ is said to be a solution of the Cauchy problem

$$
(2.8) \qquad \begin{aligned} u'(t) + Tu(t) &= 0, \ t > 0, \\ u(0) &= u_0, \end{aligned}
$$

if $u(t) \in \operatorname{dom} T$ for all $t > 0$, u is differentiable and $u'(t) = -Tu(t)$ for all $t > 0$, u in strongly continuous at 0 and $u(0) = u_0$. If T is a Hille-Yosida operator and $u_0 \in \operatorname{dom} T$, then the Cauchy problem (2.8) has a unique solution

$$
(2.9) \qquad u(t) = W(t)u_0, \ t \geq 0,
$$

where W is the stronlgy continuous semigroup generated by -T.

The asymptotic behavior of the solution u(t) as $t \to \infty$ is related to the type of the semigroup, as well as to the spectral bound s(-T) of the generator,

$$
(2.10) \qquad s(-T) := \sup\{\operatorname{Re} \lambda : \lambda \in \sigma(-T)\}.
$$

In general, the spectral bound does not belong to the spectrum of the generator. Also in general, we only have $s(-T) \leq \omega_0(W)$. We call the line $s(-T) + i\mathbb{R}$ the spectral abscissa of the generator -T, and the part of the spectrum of -T on this line the peripheral spectrum of the generator -T, $\sigma_{per}(-T) = \sigma(T) \cap (s(-T) + i\mathbb{R})$.

We are interested in finding a solution of the abstract Cauchy problem for the linear transport equation in the Banach space $X = L^1(\Omega \times S)$, when Ω is a bounded open convex subset of $\mathbb{R}^3$ or $\mathbb{R}^3$ itself. We call an element $f \in L^1(\Omega \times S)$ positive, if f is non-negative almost everywhere, and strictly positive, f is positive almost everywhere. The set of all positive elements is the positive cone $X_+ = L_+^1(\Omega \times S)$. Then, every real element $f \in L^1(\Omega \times S)$ can be written on the difference of two positive elements, and every complex element f as $f = \mathrm{Re}\, f + i\, \mathrm{Im}\, f$, where $\mathrm{Re}\, f$ and $\mathrm{Im}\, f$ are real elements. If f is an element of the positive cone, then

$$(2.11) \qquad \|f\|_1 = \int_\Omega \int_S f(x,\xi)\,dx\,d\xi ,$$

and if f is a complex element of $L^1(\Omega \times S)$, then

$$(2.12) \qquad \|f\|_1 = \sup\{\|\cos\theta \cdot \mathrm{Re}\, f + \sin\theta \cdot \mathrm{Im}\, f\|_1 : \theta \in [0,2\pi]\}.$$

Because $L^1(\Omega \times S)$, like $L^p(\Omega \times S)$ with $1 \leq p < \infty$, has an order structure, it is called a ordered Banach space over $\mathbb{C}$, see Schaefer [1974]. A linear operator A is called positive, if it maps the positive cone into itself, i.e. $AX_+ \subseteq X_+$.

THEOREM 2.3. If the positive linear operator A is bounded on X_+ with bound K, then A is bounded on X with the same bound K.

PROOF. We have $\|Af\|_1 \leq K\|f\|_1$ for all $f \in X_+$. Let f be a real element of X, then f can be written in the form $f = f_+ - f_-$, where f_+ and f_- are elements of the positive cone, and $|f| = f_+ + f_-$. We have

$$(2.13) \qquad \begin{aligned} \|Af\|_1 &= \|Af_+ - Af_-\|_1 \leq \\ &\leq K(\|f_+\|_1 + \|f_-\|_1) = K\|f\|_1 . \end{aligned}$$

Let f be a complex element of X. Then we have

$$(2.14) \qquad \begin{aligned} \|Af\|_1 &= \sup\{\|\cos\theta \cdot A\,\mathrm{Re} f + \sin\theta \cdot A\,\mathrm{Im}\, f\|_1 \cdot \theta \in [0,2\pi]\} \leq \\ &\leq K \sup\{\|\cos \cdot \mathrm{Re}\, f + \sin\theta \cdot \mathrm{Im}\, f\|_1 \cdot \theta \in [0,2\pi]\} = \\ &= K \cdot \|f\|_1 . \end{aligned}$$

$\square$

If A is a positive operator, we write $A \geq 0$. The results of Perron and Frobenius on positive matrices in the ordered Banach space $\mathbb{C}^n$ were generalized to bounded positive operators in general ordered Banach spaces. Thus, the spectral radius $r(A)$ of a bounded positive operator always belongs to the spectrum $\sigma(A)$, and under certain additional conditions, the peripheral spectrum of the bounded positive operator A, i.e. the part of the spectrum of A on the spectral circle, $\sigma_{per}(A) = \sigma(A) \cap \{\lambda \in \mathbb{C} : |\lambda| = r(A)\}$, is cyclic. See Schaefer [1980].

Let X be an ordered Banach space. For our purpose it suffices to choose $X = L^1(\Omega \times S)$. We are interested in strongly continuous semigroups $W = [W(t): t \geq 0]$ of positive operator, i.e. $W(t) \geq 0$ for all $t \geq 0$, and the generator of these semigroups. Semigroups if this kind play an important role in time dependent linear transport theory. We cite two theorems which will be need in the sequel:

THEOREM 2.4. If $W = [W(t): t \geq 0]$ is a strongly continuous semigroup of positive operators, and if the spectrum of its generator is not empty, then $s(-T) \in \sigma(-T)$.

PROOF. See Greiner, Voigt, Wolff [1981]. □

THEOREM 2.5. If $W = [W(t): t \geq 0]$ is a strongly continuous semigroup of positive operators in an L^1-space, and if the spectrum of its generator is not empty, then $\omega_0(W) = s(-T)$.

PROOF. See Derndinger [1980]. □

Let T be a closed operator in a Banach space with dense domain. Let λ_0 be an isolated point of the spectrum of T. Then there exist three bounded operators P, D and S with the following properties:
(i) $P^2 = P$, i.e. P is a projection, and P reduces T, (ii) D is a quasinilpotent operator, i.e. $r(D) = 0$, with $DP = PD = D$. The Laurent expansion of $(T - \xi I)^{-1}$, with ξ in a neighborhood of λ_0, has the form:

$$(2.15)\quad (T-\xi I)^{-1} = -(\xi-\lambda_o)^{-1}P - \sum_{n=1}^{\infty}(\xi-\lambda_o)^{-n-1}D^n + \sum_{n=0}^{\infty}(\xi-\lambda_o)^n S^{n+1}$$

See Kato [1966, III. 6.5]. In general, λ_o is not an eigenvalue. However, if $M' := PX$ is finite-dimensional, then D is a nilpotent operator, and λ_o is an eigenvalue of T. In this case, λ_o is called an eigenvalue of finite algebraic multiplicity, and dim PX is called the algebraic multiplicity of λ_o. We recall that, if λ_o is an eigenvalue of T, the dimension of $\mathrm{Ker}(T-\lambda_o I)$ is called the geometric multiplicity of λ_o. The geometric multiplicity of an eigenvalue λ_o is always less or equal to the algebraic multiplicity, if λ_o is an isolated point of $\sigma(T)$ of finite algebraic multiplicity.

Let λ_o be an eigenvalue of finite algebraic multiplicity of the generator -T of a strongly continuous semigroup W in a Banach space X. From the spectral representation of -T we obtain that

$$(2.16)\qquad W(t) = \exp(\lambda_o t)\cdot \exp(Dt)\cdot P + Z(t)(I-P),\ t\geq 0,$$

where $Z = [Z(t):t\geq 0]$ is the strongly continuous semigroup of the generator -T restricted to the Banach space $M'' := (I-P)X$. A neutron physicist or a reactor engineer would like to have that $\lambda_o = s(-T) = \omega_o(W)$ is an isolated point in the spectrum of the generator -T of algebraic multiplicity one, the spectral projection P is positive and the type of the semigroup Z in the Banach space $M'' = (I-P)X$ should be less $\omega_o(W)$, i.e. $\omega_o(Z) < \omega_o(W)$.

In the following sections, we shall prove that the collisionless operator is a Hille-Yosida operator, and that it generates a strongly continuous semigroup. Collisions introduce a perturbation by the bounded linear operator A. The main tool here is the following theorem due to Hille and Phillips.

THEOREM 2.6. If T is a Hille-Yosida operator, and A is a bounded linear operator, then $T+A$ is again a Hille-Yosida operator.

PROOF. See Kato [1966, Section IX. 2.1] or Pazy [1974, Chapter 3, 3.1.]. □

The semigroups W_0 and W generated by $-T$ and $-(T+A)$ respectively, are related through Duhamel's integral equation

$$W(t) = W_0(t) - \int_0^t W(t-s)AW_0(s)ds, \quad t \geq 0, \tag{2.17}$$

or, equivalently,

$$W(t) = W_0(t) - \int_0^t W_0(t-s)\, AW(s)ds, \quad t \geq 0. \tag{2.18}$$

This equation can be solved by iteration; the resulting series is called Dyson-Phillips expansion,

$$W(t) = \sum_{n=0}^{\infty} W_0^{(n)}(t), \quad t \geq 0, \tag{2.19}$$

where $W_0^{(o)}(t) = W_0(t)$, and

$$W_0^{(n)}(t) = -\int_0^t W_0^{(n-1)}(t-s)AW_0(s)ds, \quad n = 1,2,3,\ldots . \tag{2.20}$$

The series (2.19) converges in the operator norm. If $\|W_0(t)\| \leq M \exp(\omega t)$ for $t \geq 0$, then $\|W(t)\| \leq M \exp((\omega + M\|A\|)t)$ for $t \geq 0$.

§ 3 COLLISIONLESS TRANSPORT OPERATOR (REACTOR PROBLEM)

Let Ω be a bounded open convex domain in R^3, and let $S \subseteq R^3$ be the velocity domain which we assume to be a ball on spherical shell in R^2, $S = \{\xi \in R^3 : v_0 \leq |\xi| \leq v_1\}$ with $0 \leq v_0 \leq v_1 \leq \infty$. We shall see that the spectral properties of the collisionless transport operator depend heavily on the lower bound v_0 whether it is zero or positive. There is no similar dependence on the upper bound v_1. In case of monoenergetic particle transport, the spherical shell S reduces to the surface of the sphere with radius $v_0 = v_1$.

For each $(x,\xi) \in \Omega \times S$ with $\xi \neq 0$, we denote by $\tau = \tau(x,\xi)$ the unique nonnegative number such that $x - \tau\xi \in \partial\Omega$, and by $\delta = \delta(x,\xi)$ the unique nonnegative number such that $x - \delta\omega \in \delta\Omega$, where $\omega = \xi/|\xi|$. We have $\omega \in S^2$. Clearly $\delta = |\xi|\tau$.

Let $C^{\infty}_{B,o}(\Omega \times S)$ be the space of all functions f that satisfy the conditions (i) supp $f \subseteq \Omega \times S_{\alpha,\beta}$ for some $\beta \geq \alpha > 0$, where $S_{\alpha,\beta} := \{\xi \in R^3 : \alpha \leq |\xi| \leq \beta\}$, and (ii) f admits a (B,ϵ)-extension to $\Omega_\epsilon \times S$ for some $\epsilon > 0$. Here, Ω_ϵ is a ϵ-neighborhood of Ω, and a (B,ϵ)-extension is a function $f_\epsilon \in C^{\infty}(\Omega_\epsilon \times S)$ whose restriction to $\Omega \times S$ coincides with f and which vanishes on each incoming ray up to a point inside Ω, i.e., for each $(x,\xi) \in \Omega \times S$, there exists an $\eta \in (0,\tau(x,\xi))$ such that $f_\epsilon(x-s\xi,\xi) = 0$ for all $s > \eta$. If $f \in C^{\infty}_{B,o}(\Omega \times S)$ is a density function, then, for each characteristic line $\{x-s\xi : s \in R\}$, where $(x,\xi) \in \Omega \times S$, there are no particles on the part $\{x-s\xi : \eta < s < \tau(x,\xi)\}$ of the characteristic line. We notice that $C^{\infty}_0(\Omega \times S) \subseteq C^{\infty}_{B,o}(\Omega \times S)$. Therefore $C^{\infty}_0(\Omega \times S)$ is a dense linear subspace for all Banach spaces $L^p(\Omega \times S)$, where $1 \leq p < \infty$.

We define the operator T_0 on $C^{\infty}_{B,o}(\Omega \times S)$ by

$$T_0 f := \frac{\partial}{\partial x} \cdot \xi f, \quad f \in C^{\infty}_{B,o}(\Omega \times S). \tag{3.1}$$

Let $\lambda \in \mathbb{C}$, then the operator $\lambda I + T_0$ maps $C^{\infty}_{B,0}(\Omega \times S)$ into itself. We shall see shortly that $\lambda I + T_0$ is a bijection of $C^{\infty}_{B,0}(\Omega \times S)$ onto itself.

Let $f \in C^{\infty}_{B,0}(\Omega \times S)$ and define $g := (\lambda I + T_0)f$. Then $g \in C^{\infty}_{B,0}(\Omega \times S)$, and

$$\lambda f(x,\xi) + \frac{\partial}{\partial x} \cdot \xi f(x,\xi) = g(x,\xi), \quad (x,\xi) \in \Omega \times S. \tag{3.2}$$

The expression $(\partial/\partial x) \cdot \xi$ is the directional derivative in the direction of the straight line $\{x + s\xi : s \in \mathbb{R}\}$ through x. At any point of this line, Equation (3.2) can be written as

$$\lambda f(x + s\xi, \xi) + \frac{\partial}{\partial s} f(x + s\xi, \xi) = g(x + s\xi, \xi), \tag{3.3}$$

or also as

$$\frac{\partial}{\partial s} [\exp(\lambda s) f(x + s\xi, \xi)] = \exp(\lambda s) g(x + s\xi, \xi). \tag{3.4}$$

By assumption, f vanishes on $\partial\Omega$ for all incoming directions. Hence, integrating both members of (3.4) over the interval $(-\tau(x,\xi), 0)$ we find

$$f(x,\xi) = \int_{-\tau(x,\xi)}^{0} \exp(\lambda s) g(x + s\xi, \xi) ds, \tag{3.5}$$

or

$$f(x,\xi) = \int_{0}^{\tau(x,\xi)} \exp(-\lambda s) g(x - s\xi, \xi) ds. \tag{3.6}$$

This result may also be written in the form

$$f(x,\xi) = |\xi|^{-1} \int_{0}^{\delta(x,\xi)} \exp(-|\xi|^{-1} \lambda s) g(x - s\omega, |\xi| \omega) ds, \tag{3.7}$$

where $\omega = \xi/|\xi|$. Our analysis shows that $\lambda I + T_o$, considered as an operator in the test space $C^{\infty}_{B,o}(\Omega \times S)$, is injective. The pre-image of a function $g \in \mathrm{im}(\lambda I + T_o)$ is given by the right-hand side of (3.5), (3.6), or (3.7).

We now define the operator R^o_λ, $\lambda \in \mathbb{C}$, in $C^{\infty}_{B,o}(\Omega \times S)$ by the expression

$$(3.8) \qquad R^o_\lambda\, g(x,\xi) := \int_0^{\tau(x,\xi)} \exp(-\lambda s) g(x-s\xi,\xi)ds, \quad (x,\xi) \in \Omega \times S,$$

for all $g \in C^{\infty}_{B,o}(\Omega \times S)$. Thus, R^o_λ coincides with $(\lambda I + T_o)^{-1}$ on $\mathrm{im}(\lambda I + T_o)$. We prove the following lemma.

LEMMA 3.1. The mapping $\lambda I + T_o : C^{\infty}_{B,o}(\Omega \times S) \to C^{\infty}_{B,o}(\Omega \times S)$ is a bijection for each $\lambda \in \mathbb{C}$, and $(\lambda I + T_o)^{-1} = R^o_\lambda$.

PROOF. Let $g \in C^{\infty}_{B,o}(\Omega \times S)$. Then for some $\epsilon > 0$, g is the restriction of a function $g_\epsilon \in C^{\infty}(\Omega_\epsilon \times S)$, and $g(x-s\xi,\xi) = 0$, if $s > \eta$, where $\eta \in (0,\tau(x,\xi))$, $(x,\xi) \in \Omega \times S$. Choose $\epsilon' \in (0,\epsilon)$. For $(x,\xi) \in \Omega_{\epsilon'} \times S$, we denote by $\tau'(x,\xi)$ the unique nonnegative number for which $x - \tau'(x,\xi)\xi \in \partial\Omega_{\epsilon'}$. The function f defined by

$$(3.9) \qquad f(x,\xi) := \int_0^{\tau'(x,\xi)} \exp(-\lambda s) g_\epsilon(x-s\xi,\xi)ds$$

is well-defined for all $(x,\xi) \in \Omega_{\epsilon'} \times S$. It coincides with $R^o_\lambda g$ on $\Omega \times S$, and its restriction to $\Omega \times S$ is an element of $C^{\infty}_{B,o}(\Omega \times S)$. One readily verifies that $(\lambda I + T_o)f = g$. Thus, $\lambda I + T_o$ is surjective. We already know that $\lambda I + T_o$ is injective. This proves the lemma. []

LEMMA 3.2. Let $\lambda \in \mathbb{C}$, $\mathrm{Re}\,\lambda > 0$. The operator $(\lambda I + T_o)^{-1}$, whose domain is the test space $C^{\infty}_{B,o}(\Omega \times S)$, is a densely defined continuous operator in $L^1(\Omega \times S)$, with

$$(3.10) \qquad \|(\lambda I + T_0)^{-1}\| \leq (\mathrm{Re}\,\lambda)^{-1}$$

PROOF. For any $g \in C^{\infty}_{B,o}(\Omega \times S)$ we have

$$(3.11) \quad (\lambda I + T_0)^{-1} g(x,\xi) = \int_0^{\tau(x,\xi)} \exp(-\lambda s) g(x - s\xi, \xi)\, ds, \quad (x,\xi) \in \Omega \times S.$$

We extend g to the cylinder of all incoming rays by putting $g(x-s\xi,\xi)=0$ for $(x,\xi) \in \Omega \times S$ and $s \geq \tau(x,\xi)$. Then we may write the integral as an integral over (o, ∞). For any $\xi \in S$, $\lambda \in \mathbf{C}$ with $\mathrm{Re}\,\lambda > 0$,

$$(3.12) \quad \int_\Omega |(\lambda I + T_0)^{-1} g(x,\xi)|\, dx \leq \int_\Omega \int_0^\infty |\exp(-\lambda s) g(x - s\xi, \xi)|\, ds\, dx =$$

$$= \int_0^\infty \exp(-s.\mathrm{Re}\,\lambda) \cdot \int_\Omega |g(x - s\xi, \xi)|\, dx\, ds =$$

$$= (\mathrm{Re}\,\lambda)^{-1} \int_\Omega |g(x,\xi)|\, dx,$$

so $\|(\lambda I + T_0)^{-1} g\|_1 \leq (\mathrm{Re}\,\lambda)^{-1} . \|g\|_1$. Because $C^{\infty}_{B,o}(\Omega \times S)$ is dense in $L^1(\Omega \times S)$, the proof is finished. □

On account of Lemma 3.2, each operator R^o_λ with $\lambda \in \mathbf{C}$, $\mathrm{Re}\,\lambda > 0$, can be extended uniquely to a continuous linear operator in $L^1(\Omega \times S)$. We denote this extension by R_λ. The image $R_\lambda g$ of any $g \in L^1(\Omega \times S)$, for $\lambda \in \mathbf{C}$ with $\mathrm{Re}\,\lambda > 0$, is given by (3.8) a.e. on $\Omega \times S$. Our objective is to prove that T_0 is closable, so we can define T as the closure of T_0. We shall call T the collisionless transport operator. First we establish another property of R_λ.

LEMMA 3.4. Let $\lambda \in \mathbf{C}$, $\mathrm{Re}\,\lambda > 0$. The operator R_λ, which is the continuous extension of R^o_λ to $L^1(\Omega \times S)$, is injective.

PROOF. See KAPER, LEKKERKERKER, HEJTMANEK [1981; Chapter 12, Lemma 12.3].

Because R_λ ($\lambda \in \mathbf{C}$, $\mathrm{Re}\,\lambda > 0$) is injective, its inverse R_λ^{-1} is well defined. The operator R_λ^{-1} is an extension of $\lambda I + T_0$. More precisely,

R_λ^{-1} is the closure of $\lambda I + T_0$. We define $T := R_\lambda^{-1} - \lambda I$, $\lambda \in \mathbf{C}$, $\mathrm{Re}\,\lambda > 0$. Then T is the closure of T_0. In particular, T does not depend on λ. Obviously, $R_\lambda = (\lambda I + T)^{-1}$, if $\lambda > 0$. Since R_λ is defined for all $g \in L^1(\Omega \times S)$ and is bounded, it follows that the half space $\{\lambda \in \mathbf{C} : \mathrm{Re}\,\lambda > 0\}$ belongs to the resolvent set of $-T$. Furthermore, $\|R_\lambda\| \leq (\mathrm{Re}\,\lambda)^{-1}$, $\lambda \in \mathbf{C}$, $\mathrm{Re}\,\lambda > 0$. Hence, T is a Hille-Yosida operator. $\square$

For the sake of completeness we formulate two theorems about the spectrum of the collisionless transport operator $-T$.

THEOREM 3.5. If $v_0 > 0$, then $\sigma(-T)$ is empty.

We postpone the proof of this and the next theorem to Section 4, because Theorem 3.5 and 3.6 are special cases, if we take $h(x,\xi) = 0$ for all $(x,\xi) \in \Omega \times S$.

In the case $v_0 = 0$ the situation is entirely different. We make use of the concept of the approximate point spectrum, for which we use the notation $AP\sigma$. The approximate point spectrum $AP\sigma(T)$ of a linear operator T in a Banach space X is defined as the set of all $\lambda \in \mathbf{C}$, for which there exists a sequence (f_n) of unit vectors in X, such that $\|(\lambda I - T)f_n\|_X \to 0$ is $n \to \infty$; cf. Halmos [1967, Chapter 8]. The set $AP\sigma(T)$ is closed.

THEOREM 3.6. If $v_0 = 0$, then $\sigma(-T) = AP\sigma(-T) = \{\lambda \in \mathbf{C} : \mathrm{Re}\,\lambda \leq 0\}$.

We consider an initial boundary value problem connected with the partial differential equation

$$\frac{\partial f}{\partial t} = -\frac{\partial}{\partial x} \cdot \xi f, \quad (x,\xi) \in \Omega \times S, \quad t > 0. \tag{3.13}$$

We look for a classical solution that satisfies the initial condition

$$f(x,\xi,0) = f_0(x,\xi), \quad (x,\xi) \in \Omega \times S, \tag{3.14}$$

and the boundary condition

$$(3.15) \qquad \xi f(x,\xi,t) = 0, \; x \in \partial\Omega, \; \xi \in S_x, \; t > 0.$$

Here, $f_o \in C^1_{B,o}(\Omega \times S)$ is given. We notice, that f_o satisfies the boundary condition. We extend f_o to the cylinder of incoming rays by defining $f_o(x-s\xi,\xi) = 0$, $(x,\xi) \in \Omega$, $s \geq \tau(x,\xi)$. We apply Monge's theory of characteristics for linear partial differential equations of order one. Thus, the solution of the initial-boundary value problem is given by

$$(3.16) \qquad f(x,\xi,t) = f_o(x-t\xi,\xi), \; (x,\xi) \in \Omega \times S, \; t \geq 0.$$

This result can be written in a more concise form. For $t \geq 0$, let $W_o(t)$ denote the shift operator in $L^1(\Omega \times S)$ defined by

$$(3.17) \qquad W_o(t)g(x,\xi) := g(x-t\xi,\xi), \; (x,\xi) \in \Omega \times S, \; t \geq 0,$$

for any $g \in L^1(\Omega \times S)$ trivially extended to $\mathbb{R}^3 \times S$. Then the solution of the initial-boundary value problem can be written in the form:

$$(3.18) \qquad f(x,\xi,t) = W_o(t)f_o(x,\xi), \; t \geq 0.$$

We notice that $W_o(t)C^\infty_{B,o}(\Omega \times S) \subseteq C^\infty_{B,o}(\Omega \times S)$ for all $t \geq 0$, and that the shift semigroup $W_o := [W_o(t) : t \geq 0]$ is a strongly continuous semigroup of positive operators in every Banach lattice $L^p(\Omega \times S), 1 \leq p < \infty$. We could define the collisionless transport operator as the generator of this strongly continuous semigroup. Because the space $C^\infty_{B,o}(\Omega \times S)$ is dense in $L^1(\Omega \times S)$, and is mapped into itself by each $W_o(t)$, $t \geq 0$, $C^\infty_{B,o}(\Omega \times S)$ is a core of the generator; cf. Reed and Simon [1979, Theorem X.49 and XI.92]. Thus, the previous definition of $-T$ coincides with the definition given now.

The norm of the operator $W_o(t)$ depends on whether $v_o > 0$ or $v_o = 0$. One verifies from the definition that if $v_o > 0$, then $\|W_o(t)\| = 1$ for all $t \in [0,t_o)$, and $\|W_o(t)\| = 0$ for all $t \in [t_o, \infty)$, where $t_o := \text{diam}\,\Omega / v_o$. Thus, the semigroup is nilpotent, and its type $\omega_o(W_o) = -\infty$. On the other hand, if $v_o = 0$, then $\|W_o(t)\| = 1$ for all $t \geq 0$. Thus, the type of the semigroup $\omega_o(W_o) = 0$, and the spectral bound of its generator $s(-T) = \omega_o(W_o)$, because W_o is a strongly continuous semigroup of positive operators in an L^1-space. The physical interpretation of these results is obvious if $v_o > 0$, then all neutrons will have leaked out of Ω through the boundary $\partial\Omega$ for $t > t_o$, but if $v_o = 0$, then there are neutrons that remain inside Ω for an arbitrary long time.

§ 4 STREAMING OPERATOR (REACTOR PROBLEM)

In the last section the collisionless transport operator $-T$ was introduced. In this section we shall study the operator $T + A_1$, where A_1 is defined by

(4.1) $$A_1 f(x,\xi) := h(x,\xi) \cdot f(x,\xi), \quad (x,\xi) \in \Omega \times S, \quad f \in L^1(\Omega \times S).$$

We assume that $h \in L_+^\infty(\Omega \times S)$, so A_1 is a bounded perturbation to T with $\|A_1\| = \|h\|_\infty$. The operator $-(T + A_1)$ is called the streaming operator (for the reactor problem). It is a closed operator in $L^1(\Omega \times S)$, with dom $(T + A_1)$ = dom T.

Let λ be an arbitrary complex number. We show that $\lambda I + T_0 + A_1$, with domain $C_{B,0}^\infty(\Omega \times S)$, is injective. Take a function $f \in C_{B,0}^\infty(\Omega \times S)$, and define $g := (\lambda I + T_0 + A_1)f$; more explicitly,

(4.2) $$(\lambda + h(x,\xi))f(x,\xi) + \frac{\partial}{\partial x} \cdot \xi f(x,\xi) = g(x,\xi), \quad (x,\xi) \in \Omega \times S.$$

The expression $(\partial/\partial x) \cdot \xi$ is the derivative in the direction of ξ along the straight line $\{x - s\xi : s \in \mathbb{R}\}$ through x. Integration along this straight line is possible for almost all $(x,\xi) \in \Omega \times S$. Thus,

(4.3) $$-\frac{\partial}{\partial s}[\exp(-\lambda s)E(x,\xi,-s)f(x-s\xi,\xi)] = \exp(-\lambda s)E(x,\xi,-s).g(x-s\xi,\xi),$$

where, for almost all $(x,\xi) \in \Omega \times S$,

(4.4) $$E(x,\xi,-s) := \exp[-\int_0^s h(x-s'\xi,\xi)ds'], \quad s \in [0,\tau(x,\xi)).$$

The expression in the right member of (4.4) is a bounded measurable function of s. Furthermore, $f(x-s\xi,\xi) = 0$ for s in a neighborhood of $\tau(x,\xi)$. Hence,

(4.5) $$f(x,\xi) = \int_0^{\tau(x,\xi)} \exp(-\lambda s)E(x,\xi,-s)g(x-s\xi,\xi)ds, \quad (x,\xi) \in \Omega \times S.$$

Thus, we have expressed f in terms of g. We conclude that $\lambda I + T_o + A_1$ is injective.

In (4.5) the function g ranges over $\mathrm{im}(\lambda I + T_o + A_1)$. In general $\mathrm{im}(\lambda I + T_o + A_1)$ does not coincide with $C^{\infty}_{B,o}(\Omega \times S)$; rather, the two spaces overlap. The following lemma gives specific information about the range of $\lambda I + T_o + A_1$.

LEMMA 4.1. For each $\lambda \in \mathbb{C}$, $\mathrm{im}(\lambda I + T_o + A_1)$ is dense in $L^1(\Omega \times S)$.

PROOF. See Kaper, Lekkerkerker, Hejtmanek [1981, Chapter 12, Lemma 12.6]. □

The expression in the right member of (4.5) is meanigful for any $g \in L^1(\Omega \times S)$; we denote it by $Q_\lambda g$. Thus, for any $g \in L^1(\Omega \times S)$, any $\lambda \in \mathbb{C}$, and almost all $(x,\xi) \in \Omega \times S$,

$$Q_\lambda g(x,\xi) = \int_0^{\tau(x,\xi)} \exp(-\lambda s) E(x,\xi,-s) g(x-s\xi,\xi) ds, \tag{4.6}$$

or, equivalently

$$Q_\lambda g(x,\xi) = v^{-1} \int_0^{\delta(x,\xi)} \exp(-\lambda v^{-1} s) E(x,\xi,-v^{-1} s) g(x-s\omega,\xi) ds, \tag{4.7}$$

where $v = |\xi|$, $\omega = \xi/|\xi|$.

The operator R_λ introduced in § 3 is obtained from the operator Q_λ introduced above by taking $h = 0$. The function $Q_\lambda g$ does not necessarily belong to $L^1(\Omega \times S)$, so Q_λ need not be an operator that maps $L^1(\Omega \times S)$ into itself. The operator Q_λ is the inverse of $\lambda I + T_o + A_1$ on $\mathrm{im}(\lambda I + T_o + A_1)$, a dense subset of $L^1(\Omega \times S)$. If it occurs that, for some $\lambda \in \mathbb{C}$, Q_λ is bounded in $L^1(\Omega \times S)$, then Q_λ is the closure of $Q_\lambda | \mathrm{im}(\lambda I + T_o + A_1)$. Because $\lambda I + T + A_1$ is the closure of $\lambda I + T_o + A_1$ in $L^1(\Omega \times S)$, and $(\lambda I + T_o + A_1)^{-1}$ coincides with $Q_\lambda | \mathrm{im}(\lambda I + T_o + A_1)$, we may conclude that, in this case, $Q_\lambda = (\lambda I + T + A_1)^{-1}$ and $\lambda \in \rho(-(T + A_1))$.

The spectrum of the streaming operator depends critically upon whether the minimal velocity v_0 is strictly positive or zero. The case $v_0 > 0$ is relatively easy and can be dealt with in one statement.

THEOREM 4.2. If $v_0 > 0$, then $\sigma(-(T+A_1))$ is empty.

PROOF. The calculation of the norm of $Q_\lambda g$ gives the estimate:

$$\|Q_\lambda g\|_1 \leq M(\lambda) \cdot \operatorname{diam}\Omega \cdot \|g\|_1, \quad g \in L^1(\Omega \times S), \tag{4.8}$$

where

$$M(\lambda) := v_0^{-1}, \quad \text{for } \lambda \in \mathbb{C},\ \operatorname{Re}\lambda \geq 0, \tag{4.9}$$
$$M(\lambda) := v_0^{-1} \cdot \exp(-(v_0^{-1} \cdot \operatorname{diam}\Omega)\operatorname{Re}\lambda), \quad \text{for } \lambda \in \mathbb{C},\ \operatorname{Re}\lambda < 0.$$

Thus, $Q_\lambda g$ is bounded in $L^1(\Omega \times S)$ for any $\lambda \in \mathbb{C}$. It follows, from the remark preceding this subsection that $\lambda \in \rho(-(T+A_1))$ for all $\lambda \in \mathbb{C}$. []

The case $v_0 = 0$ is much more delicate and requires additional assumptions about the function h. These assumptions are concerned with the behavior of h as a function of ξ for small values of $|\xi|$. We shall always assume that, in addition to $h \in L_+^\infty(\Omega \times S)$, $\lim h(x,\xi)$ exists, for all $x \in \Omega$, as $|\xi| \to 0$. We define

$$\lambda^* := \inf\{\lim_{|\xi|\to 0} h(x,\xi) : x \in \Omega\} \tag{4.10}$$

Then, $0 \leq \lambda^* \leq \|h\|_\infty$. In addition we assume both of the following assumptions are valid:

Assumption (A): For any $\epsilon > 0$, there exists an $\eta > 0$ such that $h(x,\xi) > \lambda^* - \epsilon$ for all $x \in \Omega$ and all $\xi \in S$ with $|\xi| < \eta$.

Assumption (B): For any $\epsilon > 0$, there exists a positive number ρ, a closed ball $B_0 := \{x \in R^3 : |x - x^0| \leq \rho\}$ which is entirely contained in Ω, and a positive number η such that $h(x,\xi) < \lambda^* + \varepsilon$ for all $x \in B_0$ and all $\xi \in S$ with $|\xi| < \eta$.

Although Assumption (A) is sufficient for most purposes, some estimates become cumbersome. For this reason we shall sometimes adopt the following slightly stronger assumption:

Assumption (A^*). There exists a $c > 0$ such that $h(x,\xi) > \lambda^* - c|\xi|$ for all $x \in \Omega$ and all $\xi \in S$ with $\xi \neq 0$.

THEOREM 4.3. If Assumptions (A) and (B) are fulfilled, then

$$\sigma(-(T + A_1)) = AP\sigma(-(T + A_1)) = \{\lambda \in \mathbb{C} : \operatorname{Re} \lambda \leq -\lambda^*\}.$$

PROOF. See Kaper, Lekkerkerker, Hejtmanek [1981, Chapter 12]. []

The operator T is a Hille-Yosida operator in $L^1(\Omega \times S)$, A_1 is a bounded perturbation, so from the Hille-Phillips Theorem, we conclude that $T + A_1$ is also a Hille-Yosida operator in $L^1(\Omega \times S)$. Hence, $-(T+A_1)$ is the generator of a strongly continuous semigroup $W_1 = [W_1(t) : t \in \mathbb{R}_+]$ in $L^1(\Omega \times S)$. The expression for $W_1(t)$ is found by considering the classical version of the abstract Cauchy problem

$$(4.11) \quad \frac{\partial f}{\partial t} = -\frac{\partial}{\partial x} \cdot \xi f - h(x,\xi) f(x,\xi,t), \quad (x,\xi) \in \Omega \times S, \quad t \geq 0.$$

$$f(x,\xi,0) = f_0(x,\xi), \quad (x,\xi) \in \Omega \times S,$$

$$f(x,\xi,t) = 0, \quad x \in \partial\Omega, \quad \xi \in S_x, \quad t \geq 0,$$

where $f_0 \in C^{\infty}_{B,0}(\Omega \times S)$. We remark that the boundary condition is satisfied for any $f_0 \in C^{\infty}_{B,0}(\Omega \times S)$. By Monge's method of characteristics, we can solve (4.11):

$$(4.12) \quad f(x,\xi,t) = E(x,\xi,-t) f_0(x-t\xi,\xi), \quad (x,\xi) \in \Omega \times S, \quad t \geq 0,$$

when we have extended f_0 to the cylinder of incoming cays,

$$f_0(x-s\xi,\xi) = 0, \quad \text{for } (x,\xi) \in \Omega \times S, \quad s \geq \tau(x,\xi).$$

This solution coincides with the solution of the abstract Cauchy problem, so $W_1(t)$ in $L^1(\Omega \times S)$ is given by the expression

$$(4.13) \qquad W_1(t)g(x,\xi) = E(x,\xi,-t)g(x-t\xi,\xi), \ (x,\xi) \in \Omega \times S, \ t \geq 0,$$

for any $g \in L^1(\Omega \times S)$ trivially extendet to $\mathbb{R}^3 \times S$.

We can give estimates for the upper bound $\|W_1(t)\|$.

THEOREM 4.4. (i) If Assumption (A^*) is fulfilled, then

$$(4.14) \qquad \|W_1(t)\| \leq \min\{1, \exp(c.\operatorname{diam}\Omega) \cdot \exp(-\lambda^* t)\}, \ t \geq 0.$$

(ii) If Assumption (A) if fulfilled, then, for any $\epsilon > 0$, there exists a number $M(\epsilon)$ greater than or equal to 1 such that

$$(4.15) \qquad \|W_1(t)\| \leq \min\{1, \ M(\epsilon) \ \exp((-\lambda^* + c)t)\}, \ t \geq 0.$$

PROOF. See Kaper, Lekkerkerker, Hejtmanek [1981]. □

If $v_0 > 0$, then W_1 is a nilpotent semigroup, because $W_1(t) = 0$ for $t \geq t_0$, where $t_0 := \operatorname{diam}\Omega / v_0$. Hence, the type of W_1 equals $-\infty$ and the spectrum of the generator $-(T + A_1)$ is empty.

If $v_0 = 0$, and if Assumptions (A) and (B) are satisfied, then the spectral bound of $-(T + A_1)$ is known and equal to $-\lambda^*$. Remember, that W_1 consists of positive operators and the underlying space is an L^1-space. Thus, $\omega_0(W_1) = s(-(T + A_1)) = -\lambda^*$.

§ 5 TRANSPORT OPERATOR (REACTOR PROBLEM)

We define the transport operator $-(T+A) := -(T+A_1)+A_2$, where

$$(5.1) \qquad A_2 g(x,\xi) := \int_S k(x,\xi \leftarrow \xi')g(x,\xi')d\xi', \quad (x,\xi)\in\Omega\times S.$$

We denote the integral of k over its second argument by $k_p(x,\xi')$.

$$(5.2) \qquad h_p(x,\xi') := \int_S k(x,\xi \leftarrow \xi')d\xi, \quad (x,\xi')\in\Omega\times S.$$

We assume that $h_p \in L_+^\infty(\Omega\times S)$. Then A_2 is a bounded positive operator in $L^1(\Omega\times S)$ with $\|A\|_2 = \|h_p\|_\infty$. The operator $-A := -A_1 + A_2$ is called the linear collision operator.

As we have seen in the previous section, $T+A_1$ is a Hille-Yosida operator in $L^1(\Omega\times S)$. Because A_2 is a bounded perturbation, we conclude from the Hille-Phillips theorem that $T+A$ is also a Hille-Yosida operator in $L^1(\Omega\times S)$. Hence, $-(T+A)$ is the generator of a strongly continuous semigroup in $L^1(\Omega\times S)$. We denote this semigroup by $W=[W(t) : t\in \mathbb{R}_+]$; $W(t)$ satisfies Duhamel's integral equation

$$(5.3) \qquad W(t) = W_1(t) + \int_0^t W_1(t-s)A_2W(s)ds, \quad t\in\mathbb{R}_+,$$

the solution of which is given by the Dyson-Phillips series expansion

$$(5.4) \qquad W(t) = \sum_{n=0}^{\infty} W_1^{(n)}(t), \quad t\in\mathbb{R}_+,$$

where $W_1^{(o)}(t) = W_1(t)$ and

$$(5.5) \qquad W_1^{(n)}(t) = \int_0^t W_1^{(n-1)}(t-s)A_2W_1(s)ds, \quad n = 1,2,3,\dots .$$

Because W_1 consists of positive operators and the perturbation A_2 is positive, the semigroup W consists of positive operators. Moreover, $W_1(t) \leq W(t)$ for all $t \geq 0$.

Also, we can consider the collision operator $-A$ as a bounded perturbation to the collisionless transport operator. Thus, $W(t)$ satisfies a Duhamel's integral equation similar to (5.3), the solution of which is given by a Dyson-Phillips series expansion similar to (5.4) and (5.5).

The semigroup W determines the solution of a Cauchy problem for the transport operator,

$$(5.6) \qquad \begin{aligned} f'(t) &= -(T+A)f(t), \quad t>0, \\ f(0) &= f_0 \in \operatorname{dom} T \subseteq L^1(\Omega \times S). \end{aligned}$$

The solution is uniquely determined and given by

$$(5.7) \qquad f(t) = W(t)f_0, \quad t \geq 0.$$

Our ultimate goal is the study of the asymptotic behavior of $W(t)$ as $t \to \infty$. As we mentioned already, a reactor engineer or neutron physicist usually assumes that the transport system has the following behavior: The spectral bound of the transport operator $\lambda_0 := s(-(T+A))$ is a simple (algebraic) eigenvalue with a positive spectral projection P_0 and the type of the semigroup $[(I-P_0)W(t) : t \geq 0]$ is less $\omega_0(W)$, i.e.

$$(5.8) \qquad W(t) = \exp(\lambda_0 t)P_0 + Z_0(t), \quad t \geq 0$$

with $\omega_0(Z) < \omega_0(W)$, where $Z_0 := [W(t)(I-P_0) : t \geq 0]$ is a strongly continuous semigroup acting on the Banach space $(I-P_0)L^1(\Omega \times S)$. We remark that Z_0 is in general not a semigroup of positive operator in $(I-P_0)L^1(\Omega \times S)$, even if W is a semigroup of positive operators in $L^1(\Omega \times S)$. The asymptotic behavior of W can be analyzed if one knows the type of the semigroup and the spectrum of the generator. The transport operator $-(T+A)$ is obtained if one adds the bounded perturbation A_2 to the streaming operator $-(T+A_1)$.

The spectrum of the streaming operator is empty if $v_o > 0$; it coincides with the left halfplane $\{\lambda \in \mathbb{C} : \operatorname{Re}\lambda \le -\lambda^*\}$ if $v_o = 0$ and Assumptions (A) and (B) are satisfied. In the remainder of this section we shall restrict ourselves to the case $v_o = 0$ and assume validity of Assumptions (A) and (B); most of the results carry over to the case $v_o > 0$ without essential changes, if one reads $-\infty$ for λ^*. Our discussion focuses to the part of $\sigma(-(T+A))$ that belongs to the halfphase $\{\lambda \in \mathbb{C} : \operatorname{Re}\lambda > -\lambda^*\}$. We refer to it as the asymptotic part of $\sigma(-(T+A))$ and denote it by

$$(5.9) \qquad \sigma_{as}(-(T+A)) := \sigma(-(T+A)) \cap \rho(-(T+A_1)).$$

As mentioned above, we are interested in the type of the semigroup W. As W consists of positive operators and the underlying space is an L^1-space, the type of W coincides with the spectral bound of its generator $-(T+A)$. We denote this spectral bound by λ_o

$$(5.10) \qquad \lambda_o = \sup\{\operatorname{Re}\lambda : \lambda \in \sigma(-(T+A))\}.$$

Let λ_o be an isolated point in the spectrum of the closed operator $-V$. The the resolvent of $-V$ has a Laurent series expansion in a neighborhood of λ_o, see Kato [1966, Chapter III, 6, 5]:

$$(5.11) \qquad (\lambda I + V)^{-1} = \frac{P_o}{\lambda - \lambda_o} + \sum_{m=1}^{\infty} \frac{D_o^m}{(\lambda - \lambda_o)^{m+1}} - \sum_{m=0}^{\infty} (\lambda - \lambda_o)^m S^{m+1},$$

where P_o is a projection that reduces the operator V, and D_o is a quasinilpotent operator with $D_o = D_o P_o = P_o D_o$. The points $\lambda_o \in \sigma(-V)$ is called eigenvalue of finite (algebraic) multiplicity if λ_o is an isolated point of the spectrum of $-V$, and if P_o is degenerate, i.e. $\dim \operatorname{im} P_o < \infty$. It follows that D_o is nilpotent, the second term on the right hand of (5.11) reduces to a finite sum, i.e. λ_o is a pole for the resolvent, and the principal part of the Laurent series is degenerate. It follows also that λ_o is an eigenvalue of $-V$. We notice that, if $\lambda_o \in P\sigma(-V)$, it is does not follow that λ_o is eigenvalue of finite (algebraic) multiplicity. If $\lambda_o \in P\sigma(-V)$, we call $\dim \ker(\lambda I + V)$

the (geometric) multiplicity of λ_0.

Let λ_0 be an eigenvalue of finite (algebraic) multiplicity of the operator $-V$, where V is a Hille-Yosida operator. Because D_0 reduces V, the semigroup W_0 generated by $-V$ can be written as

$$W_0(t) = \exp(\lambda_0 t)\exp(D_0 t)\cdot P_0 + Z_0(t),\quad t\geq 0, \tag{5.12}$$

where $Z_0 := [W_0(t)(I-P_0) : t\geq 0]$. If λ_0 is an eigenvalue of (algebraic) multiplicity one, also called a simple (algebraic) eigenvalue, then formula (5.12) reduces to (5.8). We call λ_0 a strictly dominant eigenvalue, if $\lambda_0 = s(-V)$ is a simple (algebraic) eigenvalue with a positive eigenprojection P_0 and if there exists an $\epsilon > 0$ such that the real part of any other point of the spectrum is less than $\lambda_0 - \epsilon$. Neutron physicists call λ_0 the decay constant and the corresponding positive eigenfunction the fundamental mode of the transport system. A transport system whose transport operator has a strictly dominant eigenvalue is called critical if $\lambda_0 = 0$, subcritical if $\lambda_0 < 0$, supercritical if $\lambda_0 > 0$. Criticality of a transport system cannot be defined unless the transport operator has a strictly dominant eigenvalue.

We have proved in § 4 that the resolvent Q_λ of $-(T+A_1)$ exists for all $\lambda \in \mathbf{C}$ with $\operatorname{Re}\lambda > -\lambda^*$,

$$Q_\lambda := (\lambda I + T + A_1)^{-1},\quad \operatorname{Re}\lambda > -\lambda^*. \tag{5.13}$$

We know that this operator valued function is holomorphic for $\operatorname{Re}\lambda > -\lambda^*$. The resolvent S_λ of $-(T+A)$,

$$S_\lambda := (\lambda + T + A)^{-1},\quad \lambda \in \rho(-(T+A)), \tag{5.14}$$

can be expressed in terms of Q_λ and A_2. From the identity

$$\lambda I + T + A = (\lambda I + T + A_1)(I - Q_\lambda A_2) = (I - A_2 Q_\lambda)(\lambda I + T + A_1),$$

which holds for all $\lambda \in \mathbf{C}$ with $\operatorname{Re}\lambda > -\lambda^*$, we conclude that

(5.15) $$S_\lambda = (I - Q_\lambda A_2)^{-1} Q_\lambda, \quad \mathrm{Re}\,\lambda > -\lambda^*,$$

if $I - Q_\lambda A_2$ is boundedly invertible, or

(5.16) $$S_\lambda = Q_\lambda (I - A_2 Q_\lambda)^{-1}, \quad \mathrm{Re}\,\lambda > -\lambda^*$$

if $I - A_2 Q_\lambda$ is boundedly invertible. The importance of the operators $Q_\lambda A_2$ and $A_2 Q_\lambda$ was first pointed out by Vidav [1968]. They are the building blocks of integral operators whose compactness properties can be analyzed. We say that A_2 is revolvent compact relative to $-(T + A_1)$ if, for some positive integer n, $(Q_\lambda A_2)^n$ is compact for all $\lambda \in G$, where G is the component of the resolvent set $\rho(-(T + A_1))$ that contains a right halfplane.

An important tool for the investigation of the spectrum of the transport operator is provided by a theorem of Gohberg on holomorphic operator-valued functions from the complex plane to the space of compact operators. This theorem found its way into time-dependent transport theory through the work of Vidav. The following formulation is particularly suited for our purposes.

THEOREM 5.1. Let G be an open connected subset of $\mathbb{C}$, and let $[C(\lambda): \lambda \in G]$ be a holomorphic operator-valued function from G to the space of compact operators. If $I - C(\lambda)$ is boundedly invertible for some $\lambda \in G$, then $I - C(\lambda)$ is boundedly invertible for all $\lambda \in G$ except at a discrete set of points $\{\lambda_k : k = 1,2,3,\ldots\}$, each λ_k is a pole of $(I - C(\lambda))^{-1}$.

PROOF. See Gohberg [1951], Gohberg and Krein [1969], Smul'yan [1955], and Ribaric and Vidav [19..]. See also Kato [1966, Section VII.1.3].

We can apply Gohberg's theorem to a compact perturbation of a bounded linear operator. The resulting theorem is called a theorem of Weyl's type. We are interested in the application of Gohberg's theorem to a resolvent compact perturbation of a Hille-Yosida operator.

THEOREM 5.2. Let T be a Hille-Yosida operator, A a linear perturbation that is resolvent compact relative to -T. Let G be the component of the resolvent set $\rho(-T)$ that contains a right halfphase. Then $\lambda I + T + A$ is boundedly invertible for all $\lambda \in G$, except at a discrete set of points $\{\lambda_k : k = 1, 2, 3, \ldots\}$; each λ_i is a pole of $(\lambda I + T + A)^{-1}$ and an eigenvalue of $-(T + A)$ with finite (algebraic) multiplicity.

PROOF. If A is resolvent compact with $n = 1$, then $\|(\lambda I + T)^{-1} A\| \to 0$, as $\operatorname{Re} \lambda \to \infty$, so $I + (\lambda I + T)^{-1} A$ is certainly boundedly invertible if λ is real and sufficiently large.

If A is resolvent compact with $n = 2$, then one uses the product formula

$$(5.17) \qquad I - ((\lambda I + T)^{-1} A)^2 = (I - (\lambda I + T)^{-1} A)(I + (\lambda I + T)^{-1} A).$$

From the invertibility of $I - ((\lambda I + T)^{-1} A)^2$ follows the invertibility of each of the factors and, in particular, the invertibility of $I + (\lambda I + T)^{-1} A$. Higher powers (n) are treated similarly. □

An immediate consequence of this theorem is the following one:

THEOREM 5.3. If A_2 is resolvent compact with respect to $-(T + A_1)$, then $\sigma_{as}(-(T + A))$ consists of isolated points $\{\lambda_k : k = 1, 2, 3, \ldots\}$; each λ_k is an eigenvalue of $-(T + A)$ of finite (algebraic) multiplicity.

The proof of the last theorem was based on the interpretation of the transport operator as a perturbed streaming operator. We can reverse the argument and interprete the streamin operator as a pertubed transport operator: $-(T + A_1) = -(T + A) - A_2$. This procedure leads to additional information about the spectrum of the transport operator. In particular, we shall prove that the vertical line $\{\lambda \in \mathbb{C} : \operatorname{Re} \lambda = -\lambda^*\}$

belongs to $\sigma(-(T+A))$.

LEMMA 5.4. Let G be the component of the open set $\rho(-(T+A))$ which contains a right halfplane. If $Q_\lambda A_2$ is compact for each $\lambda \in \mathbb{C}$, $\mathrm{Re}\,\lambda > -\lambda^*$, then $S_\lambda A_2$ is compact for each $\lambda \in G$.

PROOF. From (5.15) we obtain, for each Re λ sufficiently large,

$$S_\lambda = \sum_{n=0}^{\infty} (Q_\lambda A_2)^n Q_\lambda . \tag{5.18}$$

Thus, if $Q_\lambda A_2$ is compact, then $S_\lambda A_2 = \sum_{n=0}^{\infty} (Q_\lambda A_2)^{n+1}$ is compact for Re λ sufficiently large. Because the function $[S_\lambda A_2 : \lambda \in G]$ is holomorphic, the lemma follows. □

Lemma 5.4 can be generalized to the case where some power $(Q_\lambda A_2)^n$, $n = 1,2,\ldots$, is compact for each $\lambda \in \mathbb{C}$, $\mathrm{Re}\,\lambda > -\lambda^*$. In that case, $(S_\lambda A_2)^n$ is also compact for each $\lambda \in \mathbb{C}$.

THEOREM 5.5. If A_2 is resolvent compact with respect to $-(T+A_1)$, then the vertical line $\{\lambda \in \mathbb{C} : \mathrm{Re}\,\lambda = -\lambda^*\}$ belongs to $\sigma(-(T+A))$.

PROOF. Suppose there exists a point λ' with $\mathrm{Re}\,\lambda' = -\lambda^*$ that does not belong to $\sigma(-(T+A))$. The there exists a neighborhood $V = \{\lambda \in \mathbb{C} : |\lambda - \lambda'| < \varepsilon\}$ of λ' that belongs to the component G of the open set $\rho(-(T+A))$ which contains a right halfplane. If A_2 is resolvent compact with respect to $-(T+A)$ for all $\lambda \in \mathbb{C}$, $\mathrm{Re}\,\lambda > -\lambda^*$, then A_2 is also resolvent compact with respect to $-(T+A)$ for all $\lambda \in \mathbb{C}$, $\mathrm{Re}\,\lambda > -\lambda^*$. Then, if we interprete the streaming operator as a perturbed transport operator, and apply Lemma 5.4, then we must conclude that $\lambda \in \rho(-(T+A_1))$ for all $\lambda \in V$, except for a set of isolated points. This conclusion contradicts the fact that the left halfplane $\{\lambda \in \mathbb{C} : \mathrm{Re}\,\lambda \leq -\lambda^*\}$ belongs to $\sigma(-(T+A_1))$. □

Thus, if we add the operator A_2, which is supposed to be resolvent compact with respect to $-(T+A_1)$, to the streaming operator, then the spectral abscissa of $-(T+A_1)$ remains in the spectrum of $-(T+A)$ and the asymptotic part of the spectrum of $-(T+A)$ on the right hand of the line $\{\lambda \in \mathbb{C}: \operatorname{Re}\lambda = -\lambda^*\}$ consists only of isolated points of finite (algebraic) multiplicity. It can happen that the asymptotic part of the spectrum of $-(T+A)$ is empty, also it can happen that the peripheral spectrum of $-(T+A)$, $\sigma_{per}(-(T+A)) = \sigma(-(T+A)) \cap \{\lambda \in \mathbb{C}: \operatorname{Re}\lambda = \lambda_o\}$ has more than one element. Also it can happen, that some part of the approximate spectrum of $-(T+A_1)$ on the left hand of the line $\{\lambda \in \mathbb{C}:$ $: \operatorname{Re}\lambda = -\lambda^*\}$ ceases to belong to the spectrum of $-(T+A)$. We can prove two theorems concerning the spectral bound λ_o of $-(T+A)$.

THEOREM 5.6. If A_2 is resolvent compact with respect to $-(T+A_1)$, and if $\lambda_o > -\lambda^*$, then λ_o has at least one positive eigenfunction.

PROOF. Vidav [1968]. □

THEOREM 5.7. If A_2 is resolvent compact with respect to $-(T+A_1)$, if $\lambda_o > -\lambda^*$, and if the scattering kernel in strictly positive, then λ_o is an (algebraic) simple eigenvalue with a positive eigenfunction.

PROOF. Kaper, Lekkerkerker, Hejtmanek [1981]. □

Under very general conditions an integral operator, mapping an L^1-space into itself, is weakly compact, see GOLDSTEIN [1966, Corollary III, 3.9.]: e.g. if the domain of the L^1-space has finite measure, and if the kernel of the integral operator is bounded and measurable, then the integral operator is weakly compact. The product of two weakly compact operators in an L^1-space is compact; see Dunford and Schwartz [1958, Corollary VI, 8.13]. These two facts will be used to prove the resolvent compactness of the operator A_2 with respect to $-(T+A_1)$.

Now we shall prove that $A_2 Q_\lambda A_2$ is, in fact, an integral operator in $L^1(\Omega \times S)$. We distinguish two cases, namely, a velocity domain $S = \{\xi \in R^3 : v_0 \le |\xi| \le v_1\}$ where $v_0 < v_1$, and one where $v_0 = v_1$ (monoenergetic transport). In the former case we have, for any $f \in L^1(\Omega \times S)$.

(5.19) $$A_2 Q_\lambda A_2 f(x,\xi) =$$

$$= \int_S k(x,\xi \leftarrow \xi'')[\int_0^\infty \exp(-\lambda t - \int_0^t h(x-s\xi'',\xi'')ds) \cdot (\int_S k(x-t\xi'',\xi'' \leftarrow \xi') \cdot$$

$$\cdot f(x-t\xi'',\xi')d\xi')dt]d\xi'', \quad (x,\xi) \in \Omega \times S.$$

We notice, that A_2 is a threedimensional integral, and Q_λ is a one dimensional integral. If we compose $Q_\lambda A_2$, we obtain a four dimensional integral, which is evidently less then the six dimensional integral, which we need for an integral operator over the six dimensional phase space $\Omega \times S$. If we compose $A_2 Q_\lambda A_2$, we obtain a seven dimensional integral, which suffices for an integral operator over $\Omega \times S$.

We trivially continue the functions h and k outside their domains of definition and extend the ranges of integration in the integrals over ξ' and ξ'' from S to R^3. The value of the repeated integral does not change. We introduce two new variables of integration. For fixed $t > 0$, we take $x' = x - t\xi''$, with Jacobian $dx' = t^3 d\xi''$, and interchange the order of integration:

(5.20) $$A_2 Q_\lambda A_2 f(x,\xi) =$$

$$= \int_\Omega \int_S [\int_0^\infty k(x,\xi \leftarrow \frac{x-x'}{t})k(x',\frac{x-x'}{t} \leftarrow \xi') \cdot \exp(-\lambda t -$$

$$- \int_0^t h(x-s\,\frac{x-x'}{t},\frac{x-x'}{t})ds)\frac{dt}{t^3}] \; f(x',\xi')dx'd\xi', \quad (x,\xi) \in \Omega \times S.$$

We notice that the kernel of this integral operator from $L^1(\Omega \times S)$ into itself is a one dimensional integral.

If $v_o = v_1$ (monoenergetic Transport), $A_2 Q_\lambda A_2$ is still an integral operator in $L^1(\Omega \times S)$. In this case one replaces the pair of variables (t, ξ'') in (5.19) by a new variable of integration $x' = x - t\xi''$, with Jacobian $dx' = t^2 \, dt \, d\xi''$. The result is again a representation, in a form similar to (5.20). We notice that $A_2 Q_\lambda A_2$ is a five dimensional integral, which suffices for an integral operator over the five dimensional phase space $\Omega \times S$.

The following table demonstrates that, in case of $\dim \Omega = 2$ or 3, we consist consider $A_2 Q_\lambda A_2$, whereas, in case $\dim \Omega = 1$, we can take $Q_\lambda A_2$.

$\dim \Omega = 3$	$v_o < v_1$	$v_o = v_1$
$\dim S$	3	2
$\dim \Omega \times S$	6	5
$Q_\lambda A_2$	$4(<6)$	$3(<5)$
$A_2 Q_\lambda A_2$	$7(\geq 6)$	$5(\geq 5)$
$\dim \Omega = 2$	$v_o < v_1$	$v_o = v_1$
$\dim S$	2	1
$\dim \Omega \times S$	4	3
$Q_\lambda A_2$	$3(<4)$	$2(<3)$
$A_2 Q_\lambda A_2$	$5(\geq 4)$	$3(\geq 3)$
$\dim \Omega = 1$	$v_o = v_1$	
$\dim S$	1	
$\dim \Omega \times S$	2	
$Q_\lambda A_2$	$2(\geq 2)$	
$A_2 Q_\lambda A_2$	$3(\geq 2)$	

We notice that, in the case of monoenergetic neutron transport in a slab with azimuthal symmetry, which was first studied by Lehner and Wing [1955], $Q_\lambda A_2$ is already an integral operator over the phase space $(-a, a) \times [-1, 1]$. Although $Q_\lambda A_2$ is an integral operator $A_2 Q_\lambda A_2$ is more approximate, because the kernel is the exponential integral function $E(\lambda(x-x'))$. This integral operator is a Hilbert-Schmidt operator in the Hilbert space $L^2((-a,a)\times[-1,1])$, and consequently a compact operator. It was proved by Suhadolc [1971], that this integral operator is weakly compact, as considered as an operator in the Banach space $L^1((-a,a)\times[-1,1])$.

We remark that the Rellich-Kondrachov theorem can be applied to find sufficient conditions for the compactness of the integral operator $A_2 Q_\lambda A_2$. For the proof of this theorem, see Adams [1975, Theorem 6.2, page 144]. Let us assume that we can formulate sufficient conditions for the collision density h and the scattering kernel k, such that the mapping

$$(5.21) \qquad f \in L^1(\Omega \times S) \to A_2 Q_\lambda A_2 f \in W^{1,1}(\Omega \times S),$$

is bounded in the norm of the Sobolev space $W^{1,1}(\Omega \times S)$. Because the embedding of the Sobolev space $W^{1,1}(\Omega \times S)$ into $L^1(\Omega \times S)$ is compact, then also the mapping

$$(5.22) \qquad f \in L^1(\Omega \times S) \to A_2 Q_\lambda A_2 f \in L^1(\Omega \times S),$$

is compact. For sufficient conditions for h and k such that the mapping (5.21) is bounded, see Huber [1981].

We now turn to the study of the semigroup W that is generated by the transport operator. This semigroup is defined in terms of the semigroup W_1, which is generated by the streaming operator, and the perturbation A_2. Now we concentrate on the spectral properties of the bounded operators W(t). Because W(t) is closely connected with the resolvent of the transport operator, spectral properties of W(t) can often be translated into spectral properties of $-(T+A)$. Thus we shall be able to establish some additional properties of $\sigma_{as}(-(T+A))$ as well. For the sake of convenience we shall assume throughout the remainder of this section that the collision frequency h is such that Assumptions (A^*) and (B) hold, so $W_1(t)$ satisfies the estimate (4,14)

$$\|W_1(t)\| \leq \min\{1, \exp(c.\mathrm{diam}\,\Omega) \cdot \exp(-\lambda^* t)\}, \quad t > 0$$

and the type of the $\omega_0(W_1)$ of the semigroup W_1 is $-\lambda^*$.

The operator valued function $[W_1(t)A_2 : t \geq 0]$ plays a role similar to that of $[Q_\lambda A_2 : \mathrm{Re}\,\lambda > -\lambda^*]$ in the resolvent approach. Note that $[W_1(t)A_2 : t \geq 0]$ does not have the semigroup property. The significance of this function was first recognized by Vidav [1970]. Again, we shall use compactnss arguments. As might be expected, this time the compactness has to come from the operator $W_1(t)A_2$. We shall say that A_2 is semigroup compact with respect to the streaming operator $-(T + A_1)$ if, for some positive integer n, the product $\prod_{i=1}^{n} (W_1(t_1)A_2)$ is compact for each n-tuple of positive numbers $(t_1, t_2, \ldots, t_n)$. This concept will replace the concept of resolvent compactness, which played a role in our earlier discussion.

The following theorem is due to Vidav [1970]. It describes the nature of the spectrum of W(t) outside the disc with radius $\exp(-\lambda^* t)$ centred at the origin.

THEOREM 5.8. If A_2 is semigroup compact with respect to $-(T + A_1)$, in the sense that $\prod_{i=1}^{n} (W_1(t_i)A_2)$ is compact for each n-tuple of positive numbers $(t_1, t_2, \ldots, t_n)$, and if the function $[\prod_{i=1}^{n} (W_1(t_i)A_2) :$ $t_1 > 0, t_2 > 0, \ldots, t_n > 0]$ is continuous in the uniform operator topology, then $\sigma(W(t)) \cap \{\mu \in \mathbb{C} : |\mu| > \exp(-\lambda^* t)\}$ consists of isolated points μ_k; each μ_k is an eigenvalue of W(t) of finite (algebraic) multiplicity.

PROOF: Kaper, Lekkerkerker, Hejtmanek [1981]. □

This result can be used to obtain additional information about the spectrum of $-(T + A)$. We apply an analogue of the Spectral Mapping Theorem, which connects the spectrum of the semigroup W with the spectrum of the generator $-(T+A)$:

(5.23)
$$\exp(t\sigma(-(T+A))) \subseteq \sigma(W(t))$$
$$\exp(tP\sigma(-(T+A))) = P\sigma(W(t)) \setminus \{0\}$$
$$\exp(tAP\sigma(-T+A))) \subseteq AP\sigma(W(t)).$$

Also, the multiplicity of an eigenvalue λ_k of $-(T+A)$ is at most equal to the mulitplicity of the corresponding eigenvalue $\mu_k = \exp(t\lambda_k)$ of $W(t)$. The proofs of these statements can be found in Hille and Phillips [1957, Theorem 16.7] and Greiner [1980, Proposition 1.23].

THEOREM 5.9. If the conditions of Theorem 5.8 are satisfied, then $\sigma(-(T+A)) \cap \{\lambda \in \mathbb{C}: \mathrm{Re}\,\lambda \geq -\lambda^* + \varepsilon\}$ is finite for each $\varepsilon > 0$.

PROOF: If, for some $\varepsilon > 0$, there were infinitely many eigenfunctions of $-(T+A)$ associated with eigenvalues in $\{\lambda \in \mathbb{C}: \mathrm{Re}\,\lambda \geq -\lambda^* + \varepsilon\}$, then there would by infinitely many eigenfunctions of $W(t)$ associated with eigenvalues in the region $\{\mu \in \mathbb{C}: |\mu| \geq \exp((-\lambda^* + \varepsilon)t)$. This conclusion contradicts Theorem (5.8). []

We shall now show that $A_2W_1(t)A_2$ is an integral operator in $L^1(\Omega \times S)$, if $v_0 < v_1$. From the definition of $W_1(t)$ and A_2 we obtain

(5.24)
$$A_2W_1(t)A_2\, f(x,\xi) = \int_S k(x,\xi \leftarrow \xi'')E(x,\xi'',-t)\cdot$$
$$\cdot\left[\int_S k(x-t\xi'',\xi'' \leftarrow \xi')f(x-t\xi'',\xi')d\xi'\right]d\xi'',$$

for $(x,\xi) \in \Omega \times S$. In this expression the functions f, h and k are understood to be trivially extended to $\mathbb{R}^3 \times S$. For fixed $x \in \Omega$ and $t > 0$ we change the variable of integration ξ'' to $x' = x - t\xi''$, with the Jacobian $dx' = t^3 d\xi''$. Thus,

(5.25)
$$A_2W_1(t)A_2\, f(x,\xi) = \int_\Omega\int_S H_t(x,\xi,x',\xi')f(x',\xi')dx'\,d\xi'$$

where

$$(5.26) \quad H_t(x,\xi,x',\xi') = t^{-3}\, k(x,\xi \leftarrow \frac{x-x'}{t})k(x',\frac{x-x'}{t} \leftarrow \xi'') \cdot$$

$$\cdot \exp[-\int_0^t h(x-s\,\frac{x-x'}{t},\frac{x-x'}{t})ds],\ t>0.$$

These expressions show that $A_2W_1(t)A_2$ is indeed an integral operator in $L^1(\Omega \times S)$.

If $v_0 = v_1$, then each operator A_2 contributes an integral over the two-dimensional domain S. As $\Omega \times S$ is a five-dimensional domain, three applications of A_2 are needed to cover $\Omega \times S$. One may verify that, in this case, $A_2W_1(t_1)A_2\, W_1(t_2)A_2$ is an integral operator in $L^1(\Omega \times S)$.

The eigenvalues μ_k of the operator W(t) are isolated points of finite (algebraic) multiplicity in the spectrum of W(t). However, they may acumulate near the boundary of the disc $\{\mu \in \mathbf{C} : |\mu| \leq \exp(-\lambda^* t)\}$. We therefore exclude a small neighborhood of this boundary and consider only the (finitely many) eigenvalues outside the disc $\{\mu \in \mathbf{C} : |\mu| \leq \exp((-\lambda^* + \varepsilon)t)\}$. Each of these eigenvalues, in turn, corresponds to finitely many eigenvalues of the generator $-(T+A)$ in the right halfplane $\{\lambda \in \mathbf{C} : \operatorname{Re} \lambda \geq -\lambda^* + \varepsilon\}$. Let $\{\lambda_k : k = 0,1,\ldots,n\}$ be the set of all eigenvalues of $-(T+A)$ thus obtained. Then we can prove the theorem:

THEOREM 5.10. If the conditions of Theorem (5.8) are satisfied, then

$$(5.27) \qquad W(t) = \sum_{k=0}^{n} \exp(\lambda_k t)\, \exp(D_k t) P_k + (I-P) Z_n(t),$$

where $P = P_0 + P_1 + \ldots + P_n$ and $\|Z_n(t)\| = O\,(\exp((-\lambda^* + \varepsilon)t))$ as $t \to \infty$.

PROOF See Kaper, Lekkerkerker, Hejtmanek [1981, Theorem 12.21].

§ 6 MULTIPLE SCATTERING PROBLEM

We recall that the multiple scattering problem in linear transport theory is the following Cauchy problem in $L^1(\Omega \times S)$, where $\Omega = \mathbb{R}^3$:

$$(6.1) \quad \frac{\partial f}{\partial t} = -\frac{\partial}{\partial x} \cdot \xi f(x,\xi,t) - h(x,\xi) \cdot f(x,\xi,t) + $$
$$+ \int_S k(x,\xi \leftarrow \xi')f(x,\xi',t)d\xi', \quad (x,\xi) \in \mathbb{R}^3 \times S, \ t > 0,$$

with the initial condition $f(x,\xi,0) = f_o(x,\xi)$, $(x,\xi) \in \mathbb{R}^3 \times S$. Such a problem involves the solution of the initial value problem in the entire configuration space. A target, in which particles are scattered and possibly absorbed, occupies a compact and convex subset $D \subseteq \mathbb{R}^3$. Outside D the particles move freely along straight lines with constant velocity. Thus, the collision frequency function h and the scattering kernel k vanish for all $x \in \mathbb{R}^3 \setminus D$.

The precise formulation of a multiple scattering problem is similar to that of a reactor problem. There are, however, significant differences in the spectral properties of the various operators. Our first concern is the definition of the collisionless transport operator $-T$. We want to define T as the closure of an operator T_o defined by the expression $(\partial/\partial x) \cdot \xi$ on a suitable space of test functions.

We shall say that a C^∞-function f defined on $\Omega \times S$ is rapidly decreasing at infinity if the function $[(1+|x|^2+|\xi|^2)^k D^p f(x,\xi):$ $(x,\xi) \in \Omega \times S]$ is bounded for any nonnegative integer k and any multi-index $p = (p_1, p_2, \ldots, p_6)$. Next, we introduce the open ball $S_\eta := \{\xi \in \mathbb{R}^3 : |\xi| < \eta\}$, where η is any positive number. Now we introduce the test space $C^\infty_{\downarrow,0}(\Omega \times S)$ as the set of functions $f \in C^\infty(\Omega \times S)$ that satisfy the following two conditions:
(i) supp $f \subseteq \Omega \times (S \setminus S_\eta)$ for some $\eta > 0$ and (ii) f is rapidly decreasing at infinity. This test space is dense in $L^1(\Omega \times S)$.

In $C^{\infty}_{\downarrow,0}(\Omega \times S)$ we define the operator T_0 by the expression

(6.2) $$T_0 f = \frac{\partial}{\partial x} \cdot \xi f, \quad f \in C^{\infty}_{\downarrow,0}(\Omega \times S).$$

Obviously, T_0 maps the test space into itself. The initial value problem

(6.3) $$\frac{\partial f}{\partial t} = -T_0 f, \quad t > 0$$

$$f(x,\xi,0) = f_0(x,\xi), \quad (x,\xi) \in \Omega \times S,$$

when $f_0 \in C^{\infty}_{\downarrow,0}(\Omega \times S)$, has a unique solution,

(6.4) $$f(x,\xi,t) = f_0(x-t\xi,\xi), \quad (x,\xi) \in \Omega \times S, \quad t \geq 0.$$

Of course, this is a classical solution, and $f(t)$ belongs to the test space for each $t \geq 0$. We write (6.4) in functional form,

(6.5) $$f(t) = W_0(t) f_0, \quad t \geq 0,$$

and observe that the operator $W_0(t)$ thus introduced can be extended, first to the entire space $L^1(\Omega \times S)$, and then to all values $t \in \mathbb{R}$. We denote the operator thus extended by the same symbol $W_0(t)$, so for any $g \in L^1(\Omega \times S)$ we have

(6.6) $$W_0(t) g(x,\xi) = g(x-t\xi,\xi), \quad t \in \mathbb{R},$$

for almost all $(x,\xi) \in \Omega \times S$. We notice that $W_0(t)$ is a shift operator which is positive and norm-preserving, and $W_0 := [W_0(t) : t \in \mathbb{R}]$ is a strongly continuous group of positive and norm-preserving operators.

We define the operator $-T$ in $L^1(\Omega \times S)$ as the infinitesimal generator of the group W_0,

(6.7) $$-Tf = \lim_{\delta \to 0} \delta^{-1} [W_o(\delta) f - f], \quad f \in \text{dom } T,$$

where dom T is the set of all $f \in L^1(\Omega \times S)$ for which the limit exists. The operator T is the closure of T_o. We call -T the collisionless transport operator for the multiple scattering problem.

THEOREM 6.1. For Re $\lambda \neq 0$, the inverse $(\lambda I + T)^{-1}$ exists and is a bounded linear operator in $L^1(\Omega \times S)$. It admits the representation

(6.8) $$(\lambda I + T)^{-1} g(x,\xi) = \int_0^\infty \exp(-\lambda s) g(x - s\xi, \xi) ds, \quad \text{Re } \lambda > 0,$$

and

(6.9) $$(\lambda I + T)^{-1} g(x,\xi) = -\int_0^\infty \exp(\lambda s) g(x + s\xi, \xi) ds, \quad \text{Re } \lambda < 0,$$

for any $g \in L^1(\Omega \times S)$ and almost all $(x,\xi) \in \Omega \times S$. Furthermore, $\|(\lambda I + T)^{-1}\| \leq |\text{Re } \lambda|^{-1}$.

PROOF. See Kaper, Lekkerkerker, Hejtmanek [1981, Theorem 13.3.].

It follows from Theorem 6.1, that every $\lambda \in \mathbb{C}$ with Re $\lambda \neq 0$ belongs to the resolvent set $\rho(-T)$ of the collisionless transport operator, so the spectrum $\sigma(-T)$ is entirely confined to the imaginary axis. We recall the following definitions: λ belongs to the approximate point spectrum $AP\sigma(-T)$ if there exists a sequence (f_n) of vectors $f_n \in L^1(\Omega \times S)$ such that $\|(\lambda I + T) f_n\|_1 / \|f_n\|_1 \to 0$ as $n \to \infty$, and λ belongs to the compression spectrum $K\sigma(-T)$ if $\text{im}(\lambda I + T)$ is not dense in $L^1(\Omega \times S)$; see Halmos [1967, Chapter 8]. The approximate point spectrum and the compression spectrum may overlap.

THEOREM 6.2. The spectrum $\sigma(-T)$ of the collisionless transport operator -T coincides with the imaginary axis:

(6.10) $$\sigma(-T) = AP\sigma(-T) = i\mathbb{R}.$$

PROOF. See Kaper, Lekkerkerker, Hejtmanek [1981, Corollary 13.7].

We remark that the last result is true in both cases $v_0 = 0$ and $v_0 > 0$. Thus there is a difference in the spectrum of the collisionless transport operator in the case of the reactor problem and the multiple scattering problem.

We now turn our attention to the operator $-(T + A_1)$ which was called streaming operator in case of the reactor problem and which will be called penetration operator in case of the multiple scattering problem. If $h \in L_+^{\infty}(\Omega \times S)$, where $\Omega = R^3$, then A_1 is bounded in $L^1(\Omega \times S)$, and $\|A_1\| = \|h\|_{\infty}$.

The penetration operator $-(T + A_1)$ is a perturbation of the collisionless transport operator $-T$, the perturbation being bounded in norm by $\|h\|_{\infty}$. Because the spectrum of $-T$ coincides with the imaginary axis, the spectrum of $-(T + A_1)$ is confined to the strip $\{\lambda \in \mathbb{C} : -\|h\|_{\infty} \leq \mathrm{Re}\, \lambda \leq \|h\|_{\infty}\}$. This follows from the Hille-Yosida Theorem, applied for positive and negative values of t. A representation of the semigroup W_1 which is generated by $-(T + A_1)$ is readily found

$$(6.11) \qquad W_1(t)\, f(x,\xi) = E(x,\xi,-t)\, f(x-t\xi,\xi), \quad f \in L^1(\Omega \times S),$$

where

$$(6.12) \qquad E(x,\xi,-t) = \exp\left[-\int_0^t h(x-s\xi,\xi)ds\right], \quad t \in R,$$

for almost all $(x,\xi) \in \Omega \times S$. The group W_1 consists of posiitve operators.

THEOREM 6.3. If $v_0 > 0$, then $\sigma(-(T + A_1)) = AP\sigma(-(T + A_1)) = iR$.

PROOF. See Kaper, Lekkerkerker, Hejtmanek [1981, Theorem 13.10].

The analysis of the spectrum of the penetration operator is much more delicate if $v_0 = 0$. Again, additional assumptions about the function h are needed. As in the reactor problem, we shall require that $\lim_{|\xi|\to 0} h(x,\xi)$ exists for all $x \in \Omega$. Whereas in the reactor problem the infimum of the limiting values played an important role, the supremum is a critical parameter in the multiple scattering problem. We denote the supremum by λ^{**},

$$(6.13) \qquad \lambda^{**} := \sup \{ \lim_{|\xi|\to 0} h(x,\xi) : x \in \Omega \}.$$

Assumption (A'): For any $\epsilon > 0$, there exists an $\eta > 0$ such that $h(x,\xi) < \lambda^{**} + \epsilon$ for all $x \in D$ and all $\xi \in S$ with $|\xi| < \eta$.

Assumption (B'): For any $\epsilon > 0$, there exists a $\rho > 0$, a closed ball $B_0 := \{x \in R^3 : |x - x^0| \leq \rho\}$ which is entirely contained in the interior of D, and an η, such that $h(x,\xi) > \lambda^{**} - \epsilon$ for all $x \in B_0$ and all $\xi \in S$ with $|\xi| < \eta$.

THEOREM 6.4. If $v_0 = 0$, and if Assumptions (A') and (B') are fulfilled, then

$$(6.14) \qquad \sigma(-(T + A_1)) = AP\sigma(-(T + A_1)) = \{\lambda \in C : -\lambda^{**} \leq \operatorname{Re} \lambda \leq 0\}.$$

PROOF. See Kaper, Lekkerkerker, Hejtmanek [1981, Theorem 13.13].

We turn to the transport operator $-(T + A)$, which we consider as a perturbation of the penetration operator $-(T + A_1)$ by the operator A_2,

$$(6.15) \qquad A_2 f(x,\xi) := \int_S k(x,\xi \leftarrow \xi') f(x,\xi') d\xi', \quad f \in L^1(\Omega \times S),$$

for almost all $(x,\xi) \in \Omega \times S$. The function k is nonnegative and such that $h_p \in L_+^\infty(\Omega \times S)$, where

(6,16) $$h_p(x,\xi') := \int_S k(x,\xi \leftarrow \xi')d\xi, \quad (x,\xi') \in \Omega \times S.$$

Thus, A_2 is a bounded positive operator in $L^1(\Omega \times S)$, with $\|A_2\| = \|h_p\|_\infty$.

In a multiple scattering problem, the support of the function $k(.,\xi \leftarrow \xi')$ is contained in the target domain D for each pair $(\xi,\xi') \in S \times S$.

The transport operator is the infinitesimal generator of a group of transformations in $L^1(\Omega \times S)$, $W = [W(t): t \in R]$ say. We have Duhamel's integral equation

(6.17) $$W(t) = W_1(t) + \int_0^t W_1(t-s)A_2W(s)ds,$$

and the Dyson-Phillips expansion,

(6.18) $$W(t) = \sum_{n=0}^{\infty} W_1^{(n)}(t),$$

where $W_1^{(o)}(t) = W_1(t)$ and

(6.19) $$W_1^{(n)}(t) = \int_0^t W_1(t-s)A_2W_1^{(n-1)}(s)ds, \quad n = 1,2,3,\ldots .$$

Both (6.17) and (6.18) hold for all $t \in R$. It follows from the Dyson-Phillips expansion that W consists of positive operators for $t \geq 0$; no such conclusion is possible for $t < 0$. For each $t \geq 0$ one has the ordering $W_1(t) \leq W(t)$.

We refer to the example, given in Kaper, Lekkerkerker, Hejtmanek [1981], which shows that W(t) may be indefinite for some $t < 0$. The fact, that the solution of an abstract Cauchy problem for multiple scattering is given by a group, rather than a semigroup (as for the reactor problem), implies that the dynamics of a multiple scattering problem is reversible. From the present state one can determine the state of the system at any future as well as past time. However, because W(t) does not preserve positivity for $t < 0$, we have no guarantuee that, when we retrace the evolution of a given positive initial distribution into the past, we obtain a distribution that is positive and, therefore, physically meaningful at all times. One might say that the group W distinguishes between the past and the future. In this sense, the situation is reminiscent of the Boltzmann equation of the kinetic theory of gases, where the distinction between the past and the future is manifest in Boltzmann's H-theorem.

§ 7 SCATTERING THEORY

Mathematical scattering theory is the comparison of the solution of a Cauchy problem for the transport operator with the solution of a Cauchy problem for the collisionless transport operator. The quantity of interest is the scattering operator, also called the Heisenberg operator, which was introduced in quantum mechanics to describe the scattering of a particle in a potential field. In the case of transport theory, one may think of the scattering operator as a mapping which relates a velocity distribution function of a given collection of particles in the distant part to the velocity distribution function of the same collection of particles far into the future, when, in the meantime, this collection of particles has interacted with the nuclei in a target. The definition of the scattering operator follows the pattern set by quantum mechanics. There, however, the underlying space is a Hilbert space, where as here we are dealing with a Banach lattice. Taking this difference into account, one verifies that the following definition of the scattering operator corresponds exactly to the costumary definition given in quantum mechanics - see, for example, Reed and Simon [1979, Section XI.4].

(7.1) $$S := \lim_{t\to\infty} W_o(-t)W(t)W(t)W_o(-t),$$

where the limit is taken in the strong topology of bounded linear operators in $L^1(\Omega \times S)$. In the context of scattering theory, the group W_o is refered to as the free dynamics, the group W as the interacting dynamics. The definition of the scattering operator suggests the definition of the so-called wave operators or Møller operators,

(7.2) $$\tilde{\Omega}^- := \lim_{t\to\infty} W_o(-t)W(t),$$

(7.3) $$\Omega^+ := \lim_{t\to\infty} W(t)W_o(-t),$$

where, again, the limits are taken in the strong topology of bounded linear operators in $L^1(\Omega \times S)$. If the wave operators exist, then the scattering operator exists, and we have the identity

$$S = \tilde{\Omega}^- \Omega^+. \tag{7.4}$$

Without loss of much generality we assume that the velocity space S coincides with the entire space $\mathbb{R}^3$. Because $\Omega = \mathbb{R}^3$ in multiple scattering problems, the phase space $\Omega \times S$ coincides with $\mathbb{R}^6$.

Transport systems that satisfy the condition

$$\sup \{\|W(t)\|: t \geq 0\} < \infty \tag{7.5}$$

play an important role in the sequel. We shall call them non-proliferating.

THEOREM 7.1. A necessary and sufficient condition for the existence of Ω^+ is that the transport system be nonproliferating.

PROOF. See Kaper, Lekkerkerker, Hejtmanek [1981, Theorem 13.17].

Now we turn to the wave operator $\tilde{\Omega}^-$. Here, the analysis is some what more involved. As we shall see, the condition that the transport system be nonproliferating is still necessary, but no longer sufficient for the existence of $\tilde{\Omega}^-$. A locally decaying transport system has the property that

$$\lim_{t \to \infty} \int_K \int_S |W(t)f(x,\xi)| \, dx \, d\xi = 0 \tag{7.6}$$

for all $f \in L^1(\Omega \times S)$ and each compact set K of the configuration space $\Omega = \mathbb{R}^3$. The concept of a locally decaying system has its origin in quantum-mechanical scattering theory.

THEOREM 7.2. A necessary and sufficient condition for the existence of $\tilde{\Omega}^-$ is that the transport system be nonproliferating and locally decaying.

PROOF. See Kaper, Lekkerkerker, Hejtmanek [1981, Theorem 13.18].

In the second half of this section we shall discuss a simple example of an inverse scattering problem in linear transport theory.

Computerized tomography: In this section we discuss a simple example of an inverse scattering problem in linear transport theory. The example is taken from diagnostic medicine and is concerned with image reconstruction from projections, also called computerized tomography (CT). Image reconstruction from projections is the process of producing an image of a two-dimensional distribution of some physical property from estimates of its line integrals along a finite number of lines of known locations. (Actually, we could also discuss three-dimensional image reconstruction, but for the sake of simplicity we restrict ourselves to two dimensions; applications of two-dimensional image reconstruction are certainly more prevalent.) A typical apparatus for X-ray CT consists of a tube providing a single collimated beam of X-rays (photons) and a detector unit which contains an array of X-ray detectors. The tube and the detector unit are mounted on a ring, opposite to each other; the ring is mounted in a vertical position and can be rotated in its own plane. The patient lies in a stationary position on a horizontal table along the axis of rotation and through the center of the ring. By rotating the ring, shooting a beam of X-rays through the patient at frequent regular intervals, and detecting them on the other side, one builds up a two-dimensional X-ray projection of the patient. The brightress at a point is indicative of the total attenuation of the X-rays from the source to the detector, i.e., to the line integral of the X-ray attenuation coefficient between the source and detector positions. From these line integrals, a two-dimensional image of the X-ray attenuation coefficient in the slice of of the body is produced by CT techniques. Inasmuch as different tissues have different X-ray attenuation coefficients, boundaries of organs can be delineated and healthy tissue can be distinguished from tumors. In this way CT produces cross-sectional slices of the human body without surgical intervention.

For the purposes of our discussion we assume that the X-ray beam is monochromatic (one-group approximation), that the X-ray source and detectors are negligible in size (hence, all photons travel along the same straight line from the source to the detector), and that a photon which has been absorbed or scattered along this line never reaches the detector. In addition, we assume that the collision frequency at each point is isotropic.

When we normalize our variables such that the photon speed is equal to one, the relevant phase space is $\mathbb{R}^2 \times S^1$, where S^1 is the unit circle in $\mathbb{R}^2$ centered at the origin. The transport equation is

$$\frac{\partial f}{\partial t} = - \frac{\partial}{\partial x} \omega f - h(x) f(x,\omega,t), \quad (x,\omega) \in \mathbb{R}^2 \times S^1, \; t > 0.$$

The function f corresponds to the photon density in phase space. The collision frequency h, which depends only on the spatial variable, is to be determined; the support of h is contained in a convex compact subset D (the cross-sectional slice of the human body) of $\mathbb{R}^2$. We assume $h \in L_+^\infty(\mathbb{R}^2)$.

Scattering operator. The transport equation can be formulated in functional form as a differential equation in $L^1(\mathbb{R}^2 \times S^1)$,

$$f'(t) = -(T + A_1) f(t), \; t > 0$$

The collisionless transport operator $-T$ generates the free dynamics $W_o = [W_o(t) : t \in \mathbb{R}]$,

$$W_o(t)f(x,\omega) = f(x-t\omega,\omega), \quad f \in L^1(R^2 \times S^1),$$

for almost all $(x,\omega) \in R^2 \times S^1$. The penetration operator $-(T + A_1)$ generates the interacting dynamics $W_1 = [W_1(t) : t \in R]$.

$$W_1(t)f(x,\omega) = E(x,\omega,-t)f(x-t\omega,\omega), \quad f \in L^1(R^2 \times S^1),$$

where

$$E(x,\omega,-t) = \exp\left[-\int_0^t h(x-s\omega)ds\right], \quad t \in R,$$

for almost all $(x,\omega) \in R^2 \times S^1$.

We calculate the wave operators Ω^+ and $\tilde{\Omega}^-$ for the transport system. For any $f \in L^1(R^2 \times S^1)$ we have

$$W_1(t)W_o(-t)f = [E(x,\omega,-t)f(x,\omega) : (x,\omega) \in R^2 \times S^1]$$

and

$$W_o(-t)W_1(t)f = [E(x+t\omega,\omega,-t)f(x,\omega) : (x,\omega) \in R^2 \times S^1].$$

Both expressions converge as $t \to \infty$; the limits are

$$\Omega^+f = [\exp(-\int_{-\infty}^{0} h(x+s\omega)ds)f(x,\omega) : (x,\omega) \in R^2 \times S^1]$$

and

$$\tilde{\Omega}^-f = [\exp(-\int_0^{\infty} h(x+s\omega)ds)f(x,\omega) : (x,\omega) \in R^2 \times S^1].$$

These results correspond with the physical interpretation of the wave operators: Ω^+ moves a cloud of particles backwards along a backward ray in the free dynamics, and then forward along the same way back to the starting position in the interacting dynamics; the attenuation factor thus experienced is precisely $\exp(-\int_{-\infty}^{0} h(x+s\omega)ds)$. Similarly, $\tilde{\Omega}^-$ moves a cloud of particles forward along a forward ray in the interacting dynamics, and then backward along the same ray back to the starting position in the free dynamics; the attenuation factor is $\exp(-\int_{0}^{\infty} h(x+s\omega)ds)$.

The scattering operator S is given by $S = \tilde{\Omega}^-\Omega^+$,

$$Sf = [\exp(-\int_{-\infty}^{\infty} h(x+s\omega)ds)f(x,\omega) : (x,\omega) \in R^2 \times S^1].$$

Thus, S is a multiplicative operator; the multiplying function is strictly positive and bounded, because $h \in L_+^{\infty}(R^2 \times S^1)$, $h \neq 0$, and $\text{supp } h \subset D$, where D is compact. Consequently, S is a bijective mapping from the positive cone $L_+^1(R^2 \times S^1)$ onto itself, and

$$\|S\| \leq 1, \quad \|S^{-1}\| \leq \exp(\|h\|_{\infty} \text{ diam } D).$$

Actually, the scattering operator S is a function of the collision frequency h; to emphasize this point, we shall write S(h), instead of h. Thus, S is a function from the set $\{h \in L_+^{\infty}(R^2) : \text{supp } h \subset D\}$ into the positive cone of the Banach algebra $L(L^1(R^2 \times S^1))$. The inverse problem in scattering theory is the problem of inverting the function $h \to S(h)$. In CT, this problem amounts to reconstructing h from the values of the integrals $\int_{-\infty}^{\infty} h(x+s\omega)ds$. For a fixed $\omega \in S^1$, this integral depends only on the distance from the origin to the line through x in the direction ω; we denote this distance by t. Furthermore, we use the symbol θ to denote the point on S^1 that is obtained from ω by clockwise rotation over 90°. Then we can write the integral $\int_{-\infty}^{\infty} h(x+s\omega)ds$ as $\int_{x.\theta=t} h$, where the integration is along the line $x.\theta=t$, and the inverse problem of CT amounts to reconstructing h from the function Rh, where

$$Rh(\theta,t) = R_\theta h(t) = \int_{x.\theta=t} h$$

The solution to this problem can be found in a classical article by Radon [1917]; in fact, R is known as the two-dimensional Radon transformation.

The general n-dimensional Radon transformation, and a similar transformation known as the X-ray or Roentgen transformation, have been studied extensively; see Smith, Solmon, and Wagner [1977], Hertle [1979], and Helgason [1980].

Spectrum of the collisionless transport operator -T:

	Reactor problem	Multiple scattering problem
$v_o > 0$	$\sigma(-T) = \emptyset$	$\sigma(-T) = AP\sigma(-T) = i\mathbb{R}$
$v_o = 0$	$\sigma(-T) = AP\sigma(-T) = \{\lambda \in \mathbb{C} : \operatorname{Re}\lambda \le 0\}$	$\sigma(-T) = AP\sigma(-T) = i\mathbb{R}$

Spectrum of the streaming and penetration operator $-(T+A_1)$:

	Reactor problem	Multiple scattering problem
$v_o > 0$	$\sigma(-(T+A_1)) = \emptyset$	$\sigma(-(T+A_1)) = AP\sigma(-(T+A_1)) = i\mathbb{R}$
$v_o = 0$	$\sigma(-(T+A_1)) = AP\sigma(-(T+A_1)) = \{\lambda\in\mathbb{C} : \operatorname{Re}\lambda \le -\lambda^*\}$	$\sigma(-(T+A_1)) = AP\sigma(-(T+A_1)) = \{\lambda\in\mathbb{C} : -\lambda^{**} \le \operatorname{Re}\lambda \le 0\}$

Spectrum of the transport operator $-(T+A)$:

Reactor problem

Multiple scattering problem

$v_o > 0$

$v_o = 0$

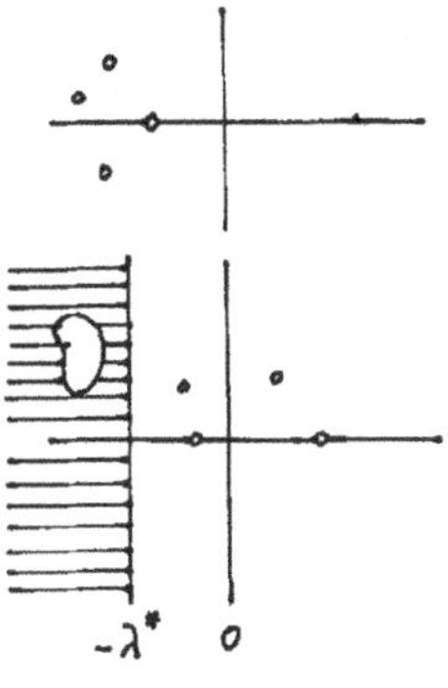

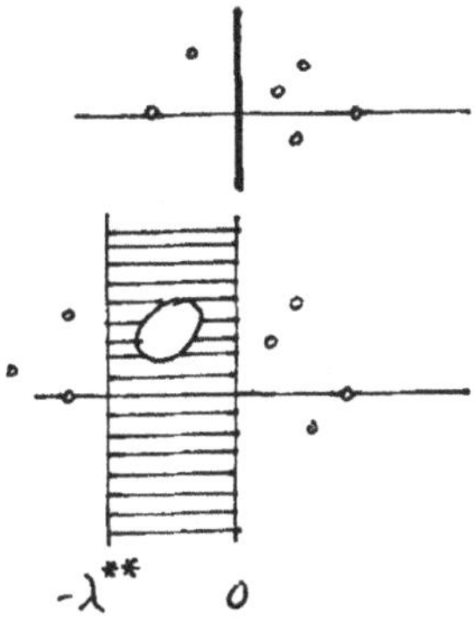

List of symbols

$-T$	collisionless transport operator
$-(T+A_1)$	streaming operator (reactor problem)
	penetration operator (multiple scattering operator)
$-(T+A)$	transport operator
$-A = -A_1 + A_2$	collision operator
$h(x,\xi)$	collision density
$k(x,\xi \leftarrow \xi')$	scattering kernel
$R_\lambda = (\lambda I + T)^{-1}$	
$Q_\lambda = (\lambda I + T + A_1)^{-1}$	
$S_\lambda = (\lambda I + T + A)^{-1}$	
$W_0(t) = \exp(-Tt)$	
$W_1(t) = \exp(-(T+A_1)t)$	
$W\ (t) = \exp(-(T+A)t)$	
$r(A)$	spectral radius of the bounded operator A
$s(-T)$	spectral bound of the generator -T of a semigroup
$\omega_0(W)$	type of the semigroup W

References

ADAMS R.A. [1975], Sobolev Spaces, Academic Press, New York (N.Y.)

DERNDINGER R. [1980], Über das Spektrum positiver Generatoren, Thesis, University of Tübingen, West Germany

DUNFORD, N. and SCHWARTZ J.T. [1958], Linear Operators, 3 Volumes, Interscience Publ., Inc., New York (N.Y.)

GOHBERG I.C. [1951], On Linear Operators Depending Analytically on a Parameter, Doklady Akad. Nauk SSSR 78, 629 - 632, (in Russian).

GOLDBERG S. [1966], Unbounded Linear Operators, Theory and Applications, McGraw-Hill Series in Higher Mathematics

GREINER G., VOIGT J., WOLFF M. [1981], On the Spectral Bound of the Generator of Semigroups of Positive Operators, J. Operator Theory 5

GREINER G. [1980], Über das Spektrum stark stetiger Halbgruppen positiver Operatoren, Thesis, University of Tübingen, West Germany

HELGASON S. [1980], The Radon Transform, Progress in Mathematics, Birkhäuser Verlag

HALMOS P.R. [1967], A Hilbert Space Problem Book,

HERTLE A. [1979], Zur Radon-Transformation von Funktionen und Maßen, Thesis, University of Erlangen-Nürnberg, West-Germany

HUBER A. [1981], Spectral Analysis of the Multiple Scattering Problem in Linear Transport Theory, Thesis, Institut für Mathematik, Universität Wien, Strudlhofgasse 4, A-1090 WIEN, Austria

KAPER H., LEKKERKERKER G., HEJTMANEK J. [1981], Spectral Methods in Linear Transport Theory, volume 3 of the series "Operator Theory, Advances and Applications", edited by I.C. Gohberg, Birkhäuser Verlag, Basel

KATO T. [1966], Perturbation Theory for Linear Operators, Springer Verlag, New York (N.Y.)

LEHNER J., WING G.M. [1955], On the Spectrum of an Asymmetric Operator Arising in the Transport Theory of Neutrons, Comm. Pure Appl. Math., 8, 217 - 234.

PAZY A. [1974], Semigroups of Linear Operators and Applications to Partial Differential Equations, Lecture Notes of the University of Maryland, # 10.

RADON J. [1917], Über die Bestimmung von Funktionen durch ihre Integralwerte längs gewisser Mannigfaltigkeiten, Sächs. Akad. Wiss. Leipzig, 69, 262 - 277.

REED M., SIMON B. [1979], Methods of Modern Mathematical Physics, 4 volumes, Academic Press, New York (N.Y.)

RIBARIC M., VIDAV I. [1969], Analytic Properties of the Inverse $A(z)^{-1}$ of an Analytic Linear Operator Valued Function A(z), Archive for Rat. Mech. and Analysis 32, 298 - 310.

SCHAEFER H.H. [1974], Banach Lattices and Positive Operators, Springer Verlag, New York (N.Y.)

SCHAEFER H.H. [1980], Ordnungsstrukturen in der Operatorentheorie Jber. d. Dt. Math.-Verein. 82, 33 - 50.

SMITH K.T., SOLMON D.C., WAGNER S.L. [1977], Practical and Mathematical Aspects of the Problem of Reconstructing Objects from Radiographs, Bulletin AMS, 83, November 1977.

SMULYAN Y. [1955], Completely Continuous Perturbation of Operators, Dokl. Akad. Nauk SSSR 101, 35 - 38 (in Russian).

SUHADOLC A. [1971], Linearized Boltzmann Equation in L^1-Space, J. Math. Anal. Appl. 35, 1 - 13.

VIDAV I. [1968], Existence and Uniqueness of Nonnegative Eigenfunctions of the Boltzmann Operator, J. Math. Anal. Appl. 22, 144 - 155.

VIDAV I. [1970], Spectra of Perturbed Semigroups with Applications to Transport Theory, J. Math. Anal. Appl. 30, 264 - 279.

These results are part of joint work together with H. Kaper, Argonne National Laboratory, and G. Lekkerkerker, University of Amsterdam, which will be published under the title "Spectral Methods in Linear Transport Theory" in Volume 3 of "Operator Theory, Advances and Applications", edited by I.C. Gohberg in Birkhäuser-Verlag.

An Introduction to the Nonlinear Boltzmann-Vlasov Equation

H. Neunzert, Universität Kaiserslautern, W.-Germany

1. The derivation of the modified Vlasov equation

The Vlasov equation in its simplest form describes the behaviour of a gas consisting of a large number of identical particles moving according to the laws of classical mechanics and interacting by a potential, which is proportional to the mass (or charge respectively) of the particles and which is - that is the main point - only weakly singular.

"Weakly singular" means precisely: The force, exerted by a particle at x on a particle at y, both of mass m, has the form $m^2 G(x,y)$, where G might become singular for x=y only in a way, that

$$\int_{\mathbb{R}^3} G(x,y)\varphi(y)dy$$

exists for all $x \in \mathbb{R}^3$, $\varphi \in L^1(\mathbb{R}^3)$.

The most typical examples are

$$(1) \quad G(x,y) = -\gamma\nabla_y \frac{1}{\|x-y\|} = -\gamma \frac{x-y}{\|x-y\|^3}$$

with $\gamma > 0$, if $m^2 G$ is the gravitational force and with $\gamma < 0$, if $m^2 G$ is the Coulomb force.

The Vlasov equation has the form

$$(2) \quad \frac{\partial f}{\partial t} + \langle v, \nabla_x f\rangle + \langle \int_{\mathbb{R}^6} G(x,y) f(t,y,v)dydv, \nabla_v f\rangle = 0$$

and is assumed to describe for G given in (1) the behaviour of a stellar gas, if $\gamma > 0$ or the behaviour of an electron gas, if $\gamma < 0$. For the stellardynamic case (2) was already considered by Jeans in 1915, while Vlasov introduced that equation for the plasmaphysical case in 1938.

Our aim is to investigate, in which sense the Vlasov equation describes the behaviour of a system of N identical particles, which move under the influence of their mutual interaction forces.

If we normalize the total mass (or charge) to 1, each particle has mass (charge) $\frac{1}{N}$ and the state is described just by giving the positions and velocities of the particles, i.e. by

$$P_i := (x_i, v_i) \in \mathbb{R}^6, \quad i=1,\dots,N.$$

The motion is then governed by Newton's law

$$(3) \qquad \begin{aligned} \dot{x}_i &= v_i \\ \dot{v}_i &= \frac{1}{N} \sum_{j=1}^{N} G(x_i, x_j). \end{aligned}$$

So, given the state $\overset{o}{\omega}_N = \{\overset{o}{P}_1, \dots, \overset{o}{P}_N\}$ at time 0, we get the state $\omega_N(t) := \{P_1(t), \dots, P_N(t)\}$ at time t by solving the initial value problem for (3) with $P_i(0) = \overset{o}{P}_i$, $i=1,\dots,N$. Such a solution might exist only for finite time - for m^2G being the gravitational force (3) represents the N-body-problem of celestial mechanics and has therefore only local solutions. Even more, the time interval of existence certainly depends on N and its length might tend to 0 if N tends to infinity.

In order to get rid of these difficulties we assume here, that G is not singular, but bounded and continuous. That can be achieved by a smooth cut-off of the potential or another mollifying procedure, for example by substituting the original G (which is weakly singular) by

$$G_\delta(x,y) := \int_{\mathbb{R}^3} G(x,z)\,\omega_\delta(z-y)\,dy$$

with the "mollifier" ω_δ defined by

$$\omega_\delta(x) := \begin{cases} 0 & \text{for } \|x\| \geq \delta \\ c_\delta \exp\left(-1 + \frac{\|x\|^2}{\delta^2}\right)^{-1} & \text{for } \|x\| < \delta \end{cases}$$

where c_δ is such that $\int_{\mathbb{R}^3} \omega_\delta \, dx = 1$.

Or by substituting the special $G(x,y) = \frac{x-y}{\|x-y\|^3}$ by

$$G_\delta(x,y) = \frac{x-y}{(\|x-y\|^2 + \delta)^{3/2}} \quad .$$

G_δ is certainly bounded and continuous; we will write again G for G_δ in that lecture. Equation (2) with G_δ instead of G is called the "modified Vlasov equation". For bounded and continuous G the initial value problem for (3) has a global solution; $\omega_N(t)$ is defined for all $t \geq 0$.

We want to compare now $\omega_N(t)$ with a function $f(t,\cdot) : \mathbb{R}^6 \to \mathbb{R}$, which is a solution of the Vlasov equation.

Let us begin with t=0: We have to compare $\omega_N(0) = \overset{\circ}{\omega}_N$ with $f(0,\cdot) = \overset{\circ}{f}$, where $\overset{\circ}{f}$ is the initial value for the Vlasov equation (2). In order to be able to compare $\overset{\circ}{\omega}_N = \{\overset{\circ}{P}_1,\ldots,\overset{\circ}{P}_N\}$ and $\overset{\circ}{f}$, we have to interprete both as mathematical objects of the same kind. Both describe mass (or charge) distributions. Mass distributions can be described by measures. Therefore interprete both - $\overset{\circ}{\omega}_N$ and $\overset{\circ}{f}$ as Borel measures.

(a) $\overset{\circ}{\omega}_N$ is substituted by the discrete measure

$$\delta_{\overset{\circ}{\omega}_N} := \frac{1}{N} \sum_{j=1}^{N} \delta_{\overset{\circ}{P}_j} ,$$

where δ_P is the usual Dirac measure concentrated on P. $\delta_{\overset{\circ}{\omega}_N}$ has total mass 1; therefore it is called a probability measure.

(b) $\overset{\circ}{f}$ is supposed to be a positive function on $\mathbb{R}^6$ with

$$\int_{\mathbb{R}^6} \overset{\circ}{f}\, dx\, dv = 1 .$$

Therefore $\overset{\circ}{f}$ might be interpreted as the density (with respect to the normal 6-dimensional Lebesgue measure) of a probability measure $\overset{\circ}{\mu}$:

$$\overset{\circ}{\mu}(M) := \int_M \overset{\circ}{f}\, dxdv \quad \text{for every Borel set } M.$$

So instead of using $\overset{\circ}{\omega}_N$ to describe the state of the N-particle system and $\overset{\circ}{f}$ to describe the initial condition for (2), we use the probability measures $\delta_{\overset{\circ}{\omega}_N}$ and $\overset{\circ}{\mu}$. We want to mention, that in spite of having probability measures there is no stochastic element in the description - we have only normalized the total mass.
We write M for the set of all probability measures on $\mathbb{R}^6$. The best method to compare $\delta_{\overset{\circ}{\omega}_N}$ and $\overset{\circ}{\mu}$ will be provided by a metric in M.

So we have to look for a metric in M. But there is one condition such a metric should fulfill. There is a natural topology in M, the weak*-topology, which I describe by the corresponding weak convergence in M:

$(\mu_n)_{N \in \mathbb{N}}$ converges weakly to μ, $\mu_n \Longrightarrow \mu$, if and only if

$$\lim_{n\to\infty} \int_{\mathbb{R}^6} \varphi \, d\mu_n = \int_{\mathbb{R}^6} \varphi \, d\mu \quad \text{for all bounded continuous } \varphi.$$

Now it is important, that a metric ρ generates the weak convergence:

$$\lim_{n\to\infty} \rho(\mu_n,\mu) = 0 \quad \text{if and only if } \mu_n \Longrightarrow \mu.$$

We have several choice for ρ, but not every choice is convenient for our purposes.

(I) There is the so-called "bounded Lipschitz distance" ρ_1 defined by

(4) $$\rho(\mu,\nu) := \sup_{\varphi\in D} \left|\int\varphi d\mu - \int \varphi \, d\nu\right|$$

where $D = \{ \varphi \,|\, \mathbb{R}^6 \longrightarrow [0,1], \ |\varphi(P)-\varphi(Q)| \le \|P-Q\|$ for all $P,Q \in \mathbb{R}^6\}$.

(II) Dobrushin and Braun-Hepp used the Wasserstein-metric $\tilde{\rho}$, defined as follows:
Consider the class Σ of measures σ on $\mathbb{R}^{12}$, such that $\sigma(M \times \mathbb{R}^6) = \mu(M)$, $\sigma(\mathbb{R}^6 \times N) = \nu(N)$. Then

(5) $$\tilde{\rho}(\mu,\nu) = \inf_{\sigma\in\Sigma} \int_{\mathbb{R}^{12}} \min\{\|P-Q\|, 1\} \, d\sigma(P,Q) .$$

(III) The most simple possible choice is the "discrepancy", defined by

(6) $$D(\mu,\nu) := \sup_{R\in\mathfrak{R}} |\mu(R)-\nu(R)|$$

where $\mathfrak{R}$ is the class of all subsets of R_6 of the form $\{P \in \mathbb{R}^6 \,|\, P \le Q\}$ with arbitrary but given Q and the usual semiorder in $\mathbb{R}^6$.

But while ρ_1, ρ_2 generate the weak convergence, this is not true for D.

Only for absolutely continuous measure μ is it true, that

$$\mu_n \Longrightarrow \mu \Longleftrightarrow D(\mu_n,\mu) \longrightarrow 0 .$$

Therefore we have to be careful in using the discrepancy and we prefer ρ_1 to work with.

Now suppose we have a sequence of N-particle systems $\overset{o}{\omega}_N$ with N increasing and an initial distribution $\overset{o}{\mu}$, so that

$$\delta_{\overset{o}{\omega}_N} \rightarrow \overset{o}{\mu} \quad \text{with } N \rightarrow \infty, \text{ or, which is the same}$$

$$\lim_{N\to\infty} \rho_1(\delta_{\overset{o}{\omega}_N}, \overset{o}{\mu}) = 0.$$

We solve, for each N, the initial value problem for (3) with

$$P_i(0) = \overset{o}{P}{}_i^N \;, \quad \text{where } \overset{o}{\omega}_N = \{\overset{o}{P}{}_1^N, \ldots, \overset{o}{P}{}_N^N\}$$

and get $\omega_N(t) = \{P_1^N(t), \ldots, P_N^N(t)\}$ or correspondingly $\delta_{\omega_N(t)}$.

On the other hand, let us assume, that $f(t,\cdot)$ is a solution of the initial value problem for (2) with $f(0,\cdot) = \overset{o}{f}$, where $\overset{o}{f}$ is the density of $\overset{o}{\mu}$. (We return to the question of existence and uniqueness of the solution of that problem later.)

It follows immediately from (2), that

$$\int_{\mathbb{R}^6} f(t,P)\,dP = \int_{\mathbb{R}^6} \overset{o}{f}(P)\,dP = 1$$

and $f(t,P) \geq 0$ for all $t \geq 0$. So $f(t,\cdot)$ is again the density of a probability measure μ_t.

We are now able to formulate the notion "derivation of the Vlasov equation".

<u>Definition 1</u>: We call the Vlasov equation (2) "strictly derivable in $[0,T]$", if for every solution μ_t of the initial value problem for (2) with initial condition $\overset{o}{\mu}$

$$\lim_{N\to\infty} \rho_1(\delta_{\omega_N(t)}, \mu_t) = 0 \quad \text{holds, if}$$

$$\lim_{N\to\infty} \rho_1(\delta_{\overset{o}{\omega}_N}, \overset{o}{\mu}) = 0.$$

<u>Remark</u>: It is clear that the definition depends on what we call a solution. If we allow a solution to be a weak solution, the definition is even stronger. On the other hand, the strong derivability

of the Vlasov equation with respect to a certain class of solutions gives you also a hint what kind of solutions are physically meaningful.

Now the question is: What conditions should be fullfilled by G, so that (2) is strictly derivable? The answer is

Theorem 1: If G is bounded, globally Lipschitz continuous, then (2) is strongly derivable (with respect to a class of very weak "measure solutions", which will be defined in the following sketch of the proof).

Sketch of the proof : We put things in a more general setting. Let

$$\mu_{\bullet} : t \longrightarrow \mu_t \in M, \quad t \in [0,T]$$

be a weakly continuous measure valued function (weakly continuous means that $t \longrightarrow \int_{\mathbb{R}^6} \varphi \, d\mu_t$ is continuous for any bounded, continuous φ).

We denote the set of these functions by C_M. Let us assume, that for any $\mu_{\bullet} \in C_M$ there is defined a six-dimensional time depending vectorfield

$$(t,P) \longrightarrow V[\mu_{\bullet}](t,P) \in \mathbb{R}^6, \quad t \in [0,T], \; P \in \mathbb{R}^6 .$$

In case of (2), this vectorfield is given by

$$(7) \qquad V[\mu_{\bullet}](t,x,v) = \begin{pmatrix} v \\ \int_{\mathbb{R}^6} G(x,y) \, d\mu_t(y,v) \end{pmatrix}$$

(remember, that $\int_{\mathbb{R}^6} G(x,y)\,d\mu_t(y,v) = \int_{\mathbb{R}^6} G(x,y) f(t,y,v)\,dy dv$).

$V[\mu_{\bullet}]$ has to satisfy two conditions, the first of which reads

(I) $V[\mu_{\bullet}](t,P)$ should be continuous in t and globally Lipschitz continuous in P (with Lipschitz constant L).

This is true for V given by (7), since $\mu_{\bullet}$ is weakly continuous, $G(x,\cdot)$ is continuous for each x and $G(\cdot,y)$ is globally Lipschitz continuous.

If (I) is true, the initial value problem for the characteristic equations

(8) $\quad \dot{P} = V[\mu_{\bullet}](t,P), \; P(s) = \overset{\circ}{P}$

has a unique globally existing solution, which we denote by

$$P(t) = T_{t,s}[\mu_{\bullet}]\overset{\circ}{P} .$$

$T_{t,s}$ is bijective and $T_{s,s} = \mathrm{id}$, $T_{t,s}^{-1} = T_{s,t}$.

Now we consider the "fix point equation"

(9) $\quad \mu_t = \overset{\circ}{\mu} \circ T_{0,t}[\mu_{\bullet}] \;, \quad t \in [0,T]$

which is to interpret as

$$\mu_t(M) = \overset{\circ}{\mu}(T_{0,t}[\mu_{\bullet}]M) \quad \text{for every Borel set } M \subset \mathbb{R}^6 .$$

(9) is a measure theoretic formulation of (2), when V is given by (7) as the following lemma shows:

<u>Lemma</u>: If $\overset{\circ}{\mu}$ is absolutely continuous with density $\overset{\circ}{f}$, then any solution μ_t of (9) is absolutely continuous with density $f(t,\cdot)$, which is a weak solution (in the sense defined by Lax for conservation laws) of the initial value problem

(10) $$\frac{\partial f}{\partial t} + \mathrm{div}_P(f \cdot V) = 0$$
$$f(0,\cdot) = \overset{\circ}{f}$$

<u>Remarks</u>:

1. If V is given by (7), equation (10) is the Vlasov equation (2).

2. A weak solution of (10) is defined to be a function f with

   (i) $f(t,\cdot)$ is weakly continuous in $[0,T]$

   (ii) for all test functions $w \in C^1([0,T] \times \mathbb{R}^6)$ with $\operatorname{supp} w \subset [0,T) \times K$, K compact in $\mathbb{R}^6$,

the equation

(11) $$\int_0^T \int_{\mathbb{R}^6} f\left[\frac{\partial w}{\partial t} + \langle V, \mathrm{grad}_P w\rangle\right] dP\,dt + \int_{\mathbb{R}^6} w(0,\cdot)\overset{\circ}{f}\, dP = 0$$

holds. f is a classical solution of (10), if $\overset{\circ}{f}$ and V is sufficiently smooth.

Remarks on the proof of the Lemma: The main point is to prove that μ_t is absolutely continuous, if $\overset{\circ}{\mu}$ is. But that follows by a rather unknown statement of Rademacher from 1916, that $\mu \circ T^{-1}$ is absolutely continuous together with μ, if T is a bijective measurable mapping and T^{-1} is locally Lipschitz continuous. So putting $T = T_{0,t}[\mu_\cdot]$, $\mu = \overset{\circ}{\mu}$, the absolute continuity of μ_t follows. The rest is essentially done by direct calculation.

The lemma shows, that (9) is a reasonable generalization of (2). But (10) includes also the N-body problem given by (3):

If we put $\mu_t = \delta_{\omega_N(t)}$, then $\int\limits_{\mathbb{R}^6} G(x,y)d\delta_{\omega_N(t)} = \frac{1}{N}\sum\limits_{j=1}^{N} G(x,x_j(t))$

and $P_i(t)$ is given as the solution of

$$\dot{P} = V[\delta_{\omega_N(\cdot)}](t,P), \quad P(0) = \overset{\circ}{P}_i, \quad i=1,\dots,N$$

where V has the form (7). It follows by (8), that

$$(12) \qquad P_i(t) = T_{t,0}[\delta_{\omega_N(\cdot)}]\overset{\circ}{P}_i \, .$$

Therefore a solution of the N-body problem (3) is a solution of equation (12) and (12) now is equivalent to

$$(13) \qquad \delta_{\omega_N(t)} = \delta_{\overset{\circ}{\omega}_N} \circ T_{0,t}[\delta_{\omega_N(\cdot)}]$$

which is just (9) with $\overset{\circ}{\mu} = \delta_{\overset{\circ}{\omega}_N}$. To verify the last equivalence of (12) and (13), just calculate for an arbitrary Borel set M

$$\delta_{P_i(t)}(M) = \delta_{T_{t,0}[\delta_{\omega_N(\cdot)}]\overset{\circ}{P}_i}(M)$$

$$= \begin{cases} 1 & \text{if } T_{t,0}[\delta_{\omega_N(\cdot)}]\overset{\circ}{P}_i \in M \\ 0 & \text{else} \end{cases} = \begin{cases} 1 & \text{if } \overset{\circ}{P}_i \in T_{0,t}[\delta_{\omega_N(\cdot)}]M \\ 0 & \text{else} \end{cases}$$

$$= \delta_{\overset{\circ}{P}_i}(T_{0,t}[\cdot]M) = \delta_{\overset{\circ}{P}_i} \circ T_{0,t}[\delta_{\omega_N(\cdot)}](M).$$

Summing up all i and dividing by N gives (13).

Conclusion: Both, the discrete N-body problem as well as the initial value problem for the Vlasov equation are special cases of equation (9), which correspond to different initial distributions $\overset{o}{\mu}$ and $\delta_{\overset{o}{\omega}_N}$ respectively. Our aim is to compare the solutions of these two special cases: We have to prove, that the solution of (9) is continuously depending on the initial data $\overset{o}{\mu}$, i.e. that

$$\delta_{\omega_N(t)} \longrightarrow \mu_t \quad \text{if} \quad \delta_{\overset{o}{\omega}_N} \longrightarrow \overset{o}{\mu} \quad \text{for } N \to \infty .$$

Now we need the second condition on V.

(II) The mapping $V : \mu_{\bullet} \longrightarrow V[\mu_{\bullet}]$ for $\mu_{\bullet} \in C_M$ is Lipschitz continuous in the following sense:

(14) $$\int_{\mathbb{R}^6} \| V[\mu_{\bullet}](t,P) - V[\nu_{\bullet}](t,P) \| \, d\mu_t(P) \leq K \, \rho_1(\mu_t,\nu_t)$$

for fixed K and all $\mu_{\bullet}, \nu_{\bullet} \in C_M$.

That condition (II) is fulfilled by V given in (7):

$$\| V[\mu_{\bullet}](t,P) - V[\nu_{\bullet}](t,P) \| = \| \int_{\mathbb{R}^6} G(x,y)(d\mu_t(y,v) - d\nu_t(y,v)) \|$$

$$= 2L \cdot B \| \int_{\mathbb{R}^6} \frac{G(x,y)+B}{2LB} (d\mu_t(y,v) - d\nu_t(y,v)) \| ,$$

where B is the bound for $\|G(x,y)\|$ and L the Lipschitz constant of G (and V) (remember, that $\int_{\mathbb{R}^6} (d\mu_t - d\nu_t) = 0$, since μ_t and ν_t are probability measures).

Now $\frac{G(x,\cdot)+B}{2LB}$ is for each x in class D used in definition (4) if $L \geq 1$; therefore we get

$$\| V[\mu_{\bullet}](t,P) - V[\nu_{\bullet}](t,P) \| \leq 2LB \, \rho_1(\mu_t,\nu_t) .$$

Since the right hand side is independent of P and $\overset{o}{\mu}$ is again a probability measure, we get (14) with $K = 2LB$.

The rest of the proof of theorem 1 is as follows, where we write $\overset{o}{\nu}$ for $\delta_{\overset{o}{\omega}_N}$, ν_t for $\delta_{\omega_N(t)}$:

$$\rho_1(\mu_t,\nu_t)=\rho_1(\overset{\circ}{\mu}\circ T_{o,t}[\mu_\cdot],\overset{\circ}{\nu}\circ T_{o,t}[\nu_\cdot])$$

$$\leq \rho_1(\overset{\circ}{\mu}\circ T_{o,t}[\mu_\cdot],\overset{\circ}{\mu}\circ T_{o,t}[\nu_\cdot])+\rho_1(\overset{\circ}{\mu}\circ T_{o,t}[\nu_\cdot],\overset{\circ}{\nu}\circ T_{o,t}[\nu_\cdot])$$

The first term on the right hand side might be estimated by using definition (4).

$$\rho_1(\overset{\circ}{\mu}\circ T_{o,t}[\mu_\cdot],\overset{\circ}{\mu}\circ T_{o,t}[\nu_\cdot]) = \sup_{\varphi\in D}\left|\int_{\mathbb{R}^6}\varphi d(\overset{\circ}{\mu}\circ T_{o,t}[\mu_\cdot])-\varphi d(\overset{\circ}{\mu}\circ T_{o,t}[\nu_\cdot])\right|$$

$$=\sup_{\varphi\in D}\left|\int_{\mathbb{R}^6}(\varphi\circ T_{t,o}[\mu_\cdot]-\varphi\circ T_{t,o}[\nu_\cdot])d\mu_o\right|$$

(since $\varphi\in D$)

$$\leq \sup\int_{\mathbb{R}^6}\|T_{t,o}[\mu_\cdot]-T_{t,o}[\nu_\cdot]\|d\mu_o=:\lambda(t)$$

Now $T_{t,o}[\mu_\cdot]P$, $T_{t,o}[\nu_\cdot]P$ are solutions of the initial value problems

$\dot{P}=V[\mu_\cdot](t,P)$ and $\dot{P}=V[\nu_\cdot](t,P)$, $P(O)=P$ respectively.

Using that and condition (II) for V we get an estimate of $\lambda(t)$:

$$\lambda(t) = \int_{\mathbb{R}^6}\|T_{t,o}[\mu_\cdot]P-T_{t,o}[\nu_\cdot]P\|\,d\mu_o$$

$$=\int_{\mathbb{R}^6}\left\|\int_0^t V[\mu_\cdot](\tau,T_{\tau,o}[\mu_\cdot]P)d\tau-\int_0^t V[\nu_\cdot](\tau,T_{\tau,o}[\nu_\cdot]P)d\tau\right\|\,d\mu_o$$

$$\leq\int_{\mathbb{R}^6}\left\|\int_0^t[V[\mu_\cdot](\tau,T_{\tau,o}[\mu_\cdot]P)-V[\nu_\cdot](\tau,T_{\tau,o}[\mu_\cdot]P)]d\tau\right\|d\mu_o\;+$$

$$+\int_{\mathbb{R}^6}\left\|\int_0^t[V[\nu_\cdot](\tau,T_{\tau,o}[\mu_\cdot]P)-V[\nu_\cdot](\tau,T_{\tau,o}[\nu_\cdot]P)]d\tau\right\|\,d\mu_o$$

$$\leq\int_{\mathbb{R}^6}\left(\int_0^t\|\ldots\|\,d\tau\right)d\mu_o+\int_{\mathbb{R}^6}\left(\int_0^t\|\ldots\|d\tau\right)d\mu_o\;=\;\text{(Fubini)}$$

$$=\int_0^t\left(\int_{\mathbb{R}^6}\|V[\mu_\cdot](\tau,T_{\tau,o}[\mu_\cdot]P)-V[\nu_\cdot](\tau,T_{\tau,o}[\mu_\cdot]P)\|\,d\mu_o(P)d\tau\;+\right.$$

$$+\int_{\mathbb{R}^6}\left(\int_0^t\|V[\nu_\cdot](\tau,T_{\tau,o}[\mu_\cdot]P)-V[\nu_\cdot](\tau,T_{\tau,o}[\nu_\cdot]P)\|\,d\tau d\mu_o\right.$$

$$\leq\int_0^t\left(\int_{\mathbb{R}^6}\|V[\mu_\cdot](\tau,Q)-V[\nu_\cdot](\tau,Q)\|\,d\mu_\tau(Q)\right)d\tau\;+$$

$$+L\int_0^t\int_{\mathbb{R}^6}\|T_{\tau,o}[\mu_\cdot]P-T_{\tau,o}[\nu_\cdot]P\|\,d\mu_o(P)d\tau\;\leq\;\text{(II)}$$

$$\leq K\int_0^t \rho_1(\mu_\tau,\nu_\tau)d\tau + L\int_0^t \lambda(\tau)d\tau \ .$$

So we have

$$\lambda(t) \leq K\int_0^t \rho_1(\mu_\tau,\nu_\tau)d\tau + L\int_0^t \lambda(\tau)d\tau \quad \text{and}$$

the wellknown lemma of Gronwall provides

$$\lambda(t) \leq K\ e^{Lt}\int_0^t \rho_1(\mu_\tau,\nu_\tau)e^{-L\tau}d\tau$$

The second term in the estimation for $\rho_1(\mu_t,\nu_t)$ is more simple:

$$\rho_1(\overset{o}{\mu}\circ T_{o,t}[\nu_\cdot],\overset{o}{\nu}\circ T_{o,t}[\nu_\cdot]) = \sup_{\varphi\in D}\left|\int\varphi\circ T_{o,t}[\nu_\cdot](d\overset{o}{\mu}-d\overset{o}{\nu})\right|$$

Since $T_{o,t}[\nu_\cdot]$ is Lipschitz continuous with constant e^{Lt}, $e^{-Lt}(\varphi\circ T_{o,t}[\nu_\cdot])$ is in D and we therefore get as a bound for the second term $e^{Lt}\rho_1(\mu_o,\nu_o)$.

Putting things together we have

$$\rho_1(\mu_t,\nu_t) \leq e^{Lt}\rho_1(\mu_o,\nu_o) + Ke^{Lt}\int_0^t \rho_1(\mu_\tau,\nu_\tau)e^{-L\tau}d\tau \ .$$

Applying Gronwalls lemma again, we get

$$\rho_1(\mu_t,\nu_t) \leq e^{(K+L)t}\rho_1(\overset{o}{\mu},\overset{o}{\nu}) \ .$$

So, if $\overset{o}{\nu} = \delta_{\overset{o}{\omega}_N}$ tends to $\overset{o}{\mu}$, $\nu_t = \delta_{\omega_N(t)}$ tends to μ_t, which is the statement of the theorem.

Remarks:

1. One recognizes, that the solution of (9) depends even in a Lipschitz continuous way with Lipschitz constant $e^{(K+L)t}$ on the initial data $\overset{o}{\mu}$. But one also can realize that the boundedness and the Lipschitz continuity of G is essential: Since $K = 2LB$, both properties expressed by B and L are used.

2. Theorem 1 covers by far more than the special situation given by the Vlasov equation. Especially V may depend in a rather general way on μ_o, whereas in the Vlasov case it depends

linearly on $f(t,\cdot)$.

3. The crucial assumption is condition (II). There the integration with respect to μ_t may be substituted by the integration with respect to $\delta_{\omega_N(t)}$.

2. Existence and uniqueness for the modified Boltzmann equation

So far, we have assumed the existence of at least a measure theoretic solution of the modified Vlasov equation (2). It is now easy to establish existence and uniqueness in using the theory we have developed in lecture 1.

Theorem 2: If G is bounded and globally Lipschitz, then for any $\overset{o}{f} \in L^1$ the modified Vlasov equation (2) has a global unique solution, which is a weak solution in the sense defined in (11). The solution is classical, if $\overset{o}{f}$ and G are in addition continuously differentiable.

Again we prove a more general statement: If V fullfills conditions (I) and (II), given in lecture 1, then the measure theoretic equation (9) has a unique weakly continuous solution $\mu_\cdot$ in any interval $[0,T]$.
Theorem 2 follows from that statement, since V given by (7) fullfills conditions (I) and (II) and, due to the lemma, a solution of (9) with absolutely continuous $\overset{o}{\mu}$ gives a weak solution of (2).

Proof: In M we defined the metric ρ_1; it induces a metric in C_M, the set of all weakly continuous functions $t \longrightarrow \mu_t \in M$, $t \in [0,T]$, where T is arbitrary, but fixed:

$$d_\alpha(\mu_\cdot, \nu_\cdot) := \sup_{t\in[0,T]} \rho_1(\mu_t, \nu_t) e^{-\alpha t}$$

Here $\alpha > 0$ may be chosen freely. Since (M, ρ_1) is a complete metric space (Kellerer), the same is true for (C_M, d_α).

We want to solve the equation (9)

$$\mu_t = \mu_0 \circ T_{0,t}[\mu_\cdot]$$

To this end, we introduce the operator

$$(15) \quad A : C_M \longrightarrow C_M \text{ with } (A\mu_\cdot)(t) := \mu_0 \circ T_{0,t}[\mu_\cdot], \; t \in [0,T] .$$

Here we have to prove, that $A\mu_\cdot \in C_M$, i.e. that

$$t \longrightarrow A\mu_\cdot(t), \; t \in [0,T]$$

is weakly continuous. But this is an immediate consequence of a well known theorem, which says, that $\mu_o \circ T_{o,t}[\mu_.]$ is weakly continuous, if

$$\mu_o(\{P = \mathbb{R}^6 \,|\, T_{o,\cdot} \text{ is continuous at } (t,P)\})=1.$$

Since $T_{o,\cdot}$, considered as a function of (t,P) is everywhere continuous (it is a solution of an ordinary differential equation system), $\mu_o \circ T_{o,t}[\mu_.]$ is weakly continuous for each μ_o .

Now we simply have to show, that A is for suitable α a contractive mapping in (C_M, d_α).

First we get for arbitrary $\mu_., \nu_. \in C_M$

$$\rho_1(A\mu_.(t), A\nu_.(t)) = \rho_1(\overset{o}{\mu} \circ T_{o,t}[\mu_.], \overset{o}{\mu} \circ T_{o,t}[\nu_.])$$
$$\leq Ke^{Lt} \int_0^t \rho_1(\mu_\tau, \nu_\tau) e^{-L\tau} d\tau$$

as we already saw in the proof of theorem 1.

$$\text{Therefore}\quad d_\alpha(A\mu_., A\nu_.) \leq K \sup_{t\in[0,T]} e^{(L-\alpha)t} \int_0^t \rho_1(\mu_\tau, \nu_\tau) e^{-L\tau} d\tau$$
$$\leq K \sup_{t\in[0,T]} e^{(L-\alpha)t} \int_0^t d_\alpha(\mu_., \nu_.) e^{-(L-\alpha)\tau} d\tau$$
$$= \frac{K}{\alpha - L} d_\alpha(\mu_., \nu_.) \quad \text{for } \alpha > L$$

Choosing $\alpha = K+L+\delta$ for $\delta > 0$ yields

$$d_\alpha(A\mu_., A\nu_.) \leq \frac{1}{1+\frac{\delta}{K}} d_\alpha(\mu_. \nu_.)$$

so A is a contractive operator and from the Banach fixpoint theorem the statement of the theorem follows.

Remark: The proof of theorem 2 is of Picard-Lindelöf type. One may also make a proof of Peano type: For $\overset{o}{\mu} = \delta_{\overset{o}{\omega}_N}$ the solution $\delta_{\omega_N(t)}$ of (9) exists. One can show by using Prohorovs theorem on compactness in M, that $\{\delta_{\omega_N(t)}, n \in \mathbb{N}\}$ is relatively compact in M for each t. Therefore for each t a sequence $(n_j(t))_{j\in\mathbb{N}}$ exists, such that $\delta_{\omega_{n_j(t)}}(t)$ converges to a measure μ_t. Now one has to show, that

n_j can be chosen independently of t, so that

$$\delta_{\omega_{n_j}(t)} \rightarrow \mu_t \qquad \text{for all } t \in [0,T].$$

This μ_t gives a solution of (9). The advantage of that procedure is not very big; one can weaken a little bit condition (I). But we need that kind of arguments in order to prove the existence of a weak solution for the non-modified Vlasov equation.

Historical remarks: The first existence proof for the modified Vlasov equation was given by Batt in 1963, who also introduced the modification. He had stricter assumptions concerning $\overset{o}{f}$. The theorems 1 and 2 given here were proved by me in 1975. Braun and Hepp in 1977 as well as Dobrushin in 1978 were not aware of these results and proved some slightly weaker results using the metric ρ_2.

3. Existence and uniqueness for the unmodified Vlasov equation

The research concerning existence and uniqueness for the unmodified Vlasov equation proceeded in two different directions:

(a) One looks for weak solutions, gets existence for dimension 3, but uniqueness is not proved up till now. Existence theorems for weak solutions have been shown by Arseneev and Illner and myself.

(b) One looks for classical solutions, for which uniqueness is easy to show. But one has to make some assumption on the initial data, which are restricting. So for $\overset{\circ}{f}$ depending only on two space and two velocity variables or for f being in a certain way symmetric, one gets existence. These results are mainly due to E. Horst (1980), but Batt did the first step in that direction; also Ukai, Okabe and Wollmann showed the existence of classical solutions for the "two-dimensional" case.

Since all the proofs of these theorems are rather long and technical, I just try to give the results and some impressions of the ideas lying behind.

In both cases the idea is initially the same: In order to solve (2) with initial condition $f(0,\cdot) = \overset{\circ}{f}$ and G given (1) one first solves the Vlasov equation with a modified G_δ - we call that equation now (2_δ) - and the same initial value $\overset{\circ}{f}$; the solution of that problem, which for any $\overset{\circ}{f} \in L^1$ exists as theorem 2 shows, is denoted by f_δ. Letting δ tend to 0, one hopes to get a solution of (2).
The difference of both cases lies in the kind of convergence one has to prove: Whereas in (a), where $\overset{\circ}{f}$ may be chosen arbitrarily in L^1, one only has to prove weak convergence of f_δ, in case (b), where $\overset{\circ}{f}$ has to be continuously differentiable one has to show uniform convergence.

But there is another thing in common for both cases: One has to show, that, if

$$E_0 = \int_{\mathbb{R}^6} \|v\|^2 \overset{\circ}{f}(x,v)\,dx\,dv < \infty$$

then there exists a constant C, not depending on t and δ, so that the kinetic energy of the gas fullfills the estimate

$$(16) \qquad E_\delta(t) := \int_{\mathbb{R}^6} \|v\|^2 f_\delta(t,x,v)\,dx\,dv \leq C\ .$$

(16) is proved by showing that energy conservation holds: Let $U_\delta(x,y)$ be the potential of G_δ, i.e.

$$G_\delta(x,y) = +\nabla_x U_\delta(x-y)$$

(for example $U_\delta(x-y) := -\gamma \int_{\mathbb{R}^3} \frac{1}{\|x-z\|}\omega_\delta(z-y)\,dz$ for one kind of mollifying or $U_\delta(x-y) = \frac{-\gamma}{(\|x-y\|^2+\delta)^{1/2}}$ for the other)

and define the potential energy of the solution f_δ by

$$(17) \qquad V_\delta(t) := \int_{\mathbb{R}^6} \left(\int_{\mathbb{R}^3} U_\delta(x-y) f_\delta(t,y,v)\,dy\,dv \right) f_\delta(t,x,v)\,dx\,dv;$$

then

$$(18) \qquad E_\delta(t) + V_\delta(t) = E_0 + V_\delta(0)$$

holds for all $t \geq 0$.

(18) is rather easy to prove by using (9).

(16) follows immediately from (18) in the plasmaphysical case, where $\gamma < 0$: $U_\delta(x-y)$ is then nonnegative, so this is true for $V_\delta(t)$ and

$$E_\delta(t) = E_0 + V_\delta(0) - V_\delta(t) \leq E_0 + V_\delta(0),$$

where $V_\delta(0)$ may be bounded uniformly with respect to δ.

(16) is harder to prove in the stellardynamic case, where $\gamma > 0$. But nevertheless it is true as was shown by Horst in using some Sobolev inequalities.

Let us now present the existence theorems:

Theorem 3: If $\overset{o}{f} \in L^1$ has the following properties:

(i) There exists a M such that $0 \le \overset{o}{f}(P) \le M$ a.e. in $\mathbb{R}^6$

(ii) $\int_{\mathbb{R}^6} \|P\|^2 \overset{o}{f}(P)dP < \infty$

(iii) $\overset{o}{\rho} : x \to \int \overset{o}{f}(x,v)dv$ is essentially bounded

then a weak solution $f(t,\cdot)$ of the initial value problem for (2) exists.

Sketch of the proof:

1. step: Let $\mu^{\delta}_{\bullet}: t \to \mu^{\delta}_t$, $t \in [0,T]$ the solution of (9) with mollified G_{δ}. Then one shows, that the set

$$\{\mu^{\delta}_t | t \in [0,T],\ \delta > 0\}$$

is uniformly tight in M; that means: to every $\varepsilon > 0$ there exists a $R > 0$, such that

$$\mu^{\delta}_t(\mathbb{R}^6 \setminus K_R) < \varepsilon \quad \text{for all } t \in [0,T],\ \delta > 0;$$

here $K_R := \{P \in \mathbb{R}^6 \mid \|P\| \le R\}$.
This is shown by considering

$$h_{\delta}(t) := \int_{\mathbb{R}^6} \|x\|^2 f_{\delta}(t,x,v)dxdv;$$

using, that f_{δ} is a solution of (9) and using (16) one gets

$$0 \le h_{\delta}(t) \le C' \quad \text{independent on } \delta > 0 \text{ and } t \in [0,T].$$

Now, for any $R > 0$

$$R^2\mu^{\delta}_t(\mathbb{R}^6 \setminus K_R) = R^2 \int_{\mathbb{R}^6 \setminus K_R} f_{\delta}(t,P)dP \le \int_{\mathbb{R}^6 \setminus K_R} \|P\|^2 f_{\delta}(t,P)dP$$

$$\le \int_{\mathbb{R}^6} \|P\|^2 f_{\delta}(t,P)dP = h_{\delta}(t) + E_{\delta}(t) \le C+C'$$

again using (16); therefore $\mu^{\delta}_t(\mathbb{R}^6 \setminus K_R)$ can be made smaller than an arbitrary $\varepsilon > 0$ by choosing R large enough.

Now Prohorovs theorem shows, that for any countable dense subset $T' \in [0,T]$ there is a monotone sequence $(\delta_n)_{n \in \mathbb{N}}, \delta_n \searrow 0$, such that

$$\mu^n_t := \mu^{\delta_n}_t \to \mu_t \in M \qquad \text{for } n \to \infty,\ t \in T'.$$

We have next to show, that the convergence holds also for $t \notin T'$.

2. step: We show, that for any continuously differentiable φ with compact support, the set of functions

$$t \longrightarrow \phi_n(t) := \int_{\mathbb{R}^6} \varphi \, d\mu_t^n, \quad t \in [0,T]$$

is equicontinuous. This is a technical proof, using mainly the boundedness of $\overset{o}{f}$ and therefore also of f_δ.

Now, since ϕ_N converges pointwise on T' and forms an equicontinuous set, it converges uniformly on [0,T]. The limit has to be

$$\lim_{n \to o} \phi_n(t) = \int_{\mathbb{R}^6} \varphi \, d\mu_t$$

with $\mu_t \in M$ defined for all $t \in [0,T]$. It follows immediately, that

$$\mu_t^n \longrightarrow \mu_t \quad \text{for } t \in [0,T]$$

and that $\mu_\bullet : t \longrightarrow \mu_t$ is weakly continuous.

It is also easy to show, that μ_t is absolutely continuous with a density $f(t,\cdot)$ essentially bounded by M.

3. step: What stays is a long but straightforward proof, that $f(t,\cdot)$, constructed in step 2 is a weak solution of the initial value problem for (2). We will not go into details.

Remark: The existence proof sketched here is not constructive, since it uses a compactness argument and we are not able to show, that for any sequence $(\delta_n)_{n\in\mathbb{N}}$, $\delta_n \searrow 0$ the measures $\mu_t^{\delta_n}$ converge to μ_t. This has two consequences:

(i) We are not able to prove uniqueness - we will even not make any conjecture on that problem.

(ii) Since it is not possible to show, that $\rho_1(\mu_t^\delta,\mu_t)$ is small for small δ, we cannot prove, that the unmodified Vlasov equation is strictly derivable - even for the plasmaphysical case, where the discrete problem has a global solution. It might be true, that the equation is only stochastically derivable - a notion, which we will not define here, since it is more connected with the Boltzmann equation.

We now turn to the question of existence of classical solutions, i.e. to Horst's work. Here we will always assume that $\overset{o}{f}$ is continuously differentiable and has compact support.
As already mentioned there is up till now no existence theorem for the full 3-dimensional problem. Therefore we have to consider the following lower dimensional cases:

(A) The function f in equation (2) depends only on 2 space and 2 velocity variables. Then equation (1) has to be changed into

$$G(x,y) = \gamma \nabla_y \ln \|x-y\| = -\gamma \frac{x-y}{\|x-y\|^2}$$

(B) Let H be the group of all orthogonal transformations S of $\mathbb{R}^3$ with det $S = 1$. A function $g : (x,y) \longrightarrow \mathbb{R}$, $(x,v) \in \mathbb{R}^6$ is called to be spherically symmetric if $g(Sx,Sv) = g(x,v)$ for all $S \in H$, $(x,v) \in \mathbb{R}^6$.

Now it is easy to show, that if $\overset{o}{f}$ is spherically symmetric, a solution of the corresponding initial value problem for (2) is also spherically symmetric. So the question occurs, whether such a solution for a spherically symmetric initial condition exists.

(C) Let Z be the group of orthogonal transformations Z_δ of $\mathbb{R}^3$, which are represented by matrices of the form

$$Z_\delta := \begin{pmatrix} \cos\theta & \sin\theta & 0 \\ -\sin\theta & \cos\theta & 0 \\ 0 & 0 & 1 \end{pmatrix}, \quad 0 \leq \theta \leq 2\pi$$

A function g is called rotationally symmetric, if $g(Z_\delta x, Z_\delta) = g(x,v)$ for all $(x,v) \in \mathbb{R}^6$, $Z_\delta \in Z$. Again for $\overset{o}{f}$ rotationally symmetric the corresponding solution $f(t,\cdot)$ of (2) is also rotationally symmetric and one may ask for the existence of such solutions. This problem may be considered as a 5-dimensional version of the original 6-dimensional problem. The special part of the z-axis given by the form of Z_δ is not essential.

<u>Theorem 4</u>: In each case (A), (B) or (C) there exists a global classical solution, which is unique.

Sketch of the proof: One again starts with the solutions f_δ of the modified problem. The main tool is the following lemma, which was in a similar form first proved by Batt.

Lemma: There exists a unique classical solution of the initial value problem for (2) in [0,T], if and only if

$$(19) \qquad \sup\{ \int_{\mathbb{R}^3} f_\delta(t,x,v)dv \mid t \in [0,T],\ x \in \mathbb{R}^3,\ \delta > 0\} < \infty$$

The proof of that lemma is rather long but purely technical. One gets an estimate of the form

$$\| T_{t,\tau}[\mu_\cdot^\delta]P - T_{t,\tau}[\mu_\cdot^{\tilde\delta}]P \| \le C |\delta^{\frac{1}{4}} - \tilde\delta^{\frac{1}{4}}| \ ,$$

C depending only on the length T of the time interval.
Therefore $T_{t,\tau}[\mu_\cdot^\delta]P$ converges uniformly with respect to $t,\tau \in [0,T]$, $P \in \mathbb{R}^6$ for δ tending to zero to a mapping $T_{t,\tau}(P)$.
Defining $f(t,P) := \overset{o}{f}(T_{o,t}P)$ one gets a solution of (2).
Since we have now full convergence for δ tending to zero (in contrast to the situation with weak solutions, where we only know the existence of a sequence $(\delta_j)_{j \in \mathbb{N}}$ for which $\mu_\cdot^{\delta_j}$ and $T_{t,o}[\mu_\cdot^{\delta_j}]$ converges), uniqueness also follows immediately.

For the rest one needs again (16), i.e. the fact, that the kinetic energy of f_δ is bounded uniformly with respect to δ. Applying Hölder's inequality, one gets for

$$\rho_\delta(t,x) = \int_{\mathbb{R}^3} f_\delta(t,x,v)dv$$

that
$$\|\rho_\delta(t,\cdot)\|_{\frac{5}{3}} \le K \|\overset{o}{f}\|_\infty^{\frac{2}{5}} \cdot E_\delta^{\frac{3}{5}} \le C_1 \ .$$

But in order to use the lemma, we need a uniform bound for $\|\rho_\delta\|_\infty$. Therefore one proceeds as follows: We estimate

$$(20) \qquad K_\delta(t,x) := \int_{\mathbb{R}^3} G_\delta(x-y) f_\delta(t,y,v)dydv = \int_{\mathbb{R}^3} G_\delta(x-y)\rho_\delta(t,y)dy$$

by Sobolev's inequality to get

(21) $$\|K_\delta(t,\cdot)\|_\infty \leq C_2 \|\rho_\delta\|_\infty^{\frac{4}{9}}$$

But if we are aware of the symmetry of f_δ, for example the spherical symmetry, one gets better estimates:

(22) $$|K_\delta(t,x)| \leq C_3 \frac{1}{\sqrt{x_1^2+x_2^2}} \|\rho_\delta\|_\infty^{\frac{1}{6}} .$$

That is the point where symmetry comes in! Now recall, that $T_{t,o}[\mu^\delta_\cdot]P$ is a solution of the system

$$\dot{x} = v$$
$$\dot{v} = K_\delta(t,x).$$

Therefore estimates for E_δ give estimates for $T_{t,o}[\mu^\delta_\cdot]$. Considering only the last three components of $T_{t,o}[\mu^\delta_\cdot]$, which correspond to the velocities and estimating

$$G_j^\delta := \sup\{\,|(T_{t,o}[\mu^\delta_\cdot]P)_{3+j} - P_{3+j}|\ \ |P \in \mathbb{R}^6,\ t \in [0,T]\}$$

one gets, that if $|K_\delta(t,x)| \leq g(x_j)$ for some $j \in \{1,2,3\}$ and $g \in L^P$, then

$$G_j^\delta \leq C_4 \|g\|_p^{\frac{p}{p+1}} ,$$

C_4 only depending on p and T.
Now using (21) for G_3^δ gives

$$G_3^\delta \leq C_5 \|\rho_\delta\|_\infty^{\frac{4}{9}} .$$

Using (22) for G_1^δ and G_2^δ gives

$$G_1^\delta, G_2^\delta \leq C_6 \|\rho_\delta\|_\infty^{\frac{11}{45}} .$$

The exponents look strange but are important. Since

$$\rho_\delta(t,x) = \int_{\mathbb{R}^3} \overset{o}{f}(T_{o,t}[\mu^\delta_\cdot]P)\,dv$$

these estimates for G_j^δ can be used again to give estimates for ρ_δ. One gets

$$\|\rho_\delta\|_\infty \leq \alpha(G_1^\delta+\beta)(G_2^\delta+\beta)(G_3^\delta+\beta)$$

(α,β independent of δ) and therefore

$$(23) \quad \|\rho_\delta\|_\infty \leq O\left(\|\rho_\delta\|_\infty^{\frac{14}{15}}\right) \quad \text{uniformly in } \delta.$$

Note, that $\frac{14}{15} = \frac{4}{9} + \frac{11}{45} + \frac{11}{45}$ - and note that without symmetry we would have instead of $\frac{14}{15}$ a number larger than one: $3 \cdot \frac{4}{9} = \frac{4}{3}$.

You see, that (23) cannot be true, if $\|\rho_\delta\|_\infty$ is not bounded. This proves (19) and therefore theorem 4.

Remark: I gave some of the most important estimates in Horst's work in order to show you how tough the stuff is. The methods end up with an estimate of the type

$$\|\rho_\delta\|_\infty = O(\|\rho_\delta\|_\infty^k)$$

and if $k < 1$, then we are through. In the twodimensional case we have just $k = 2 \cdot \frac{4}{9} < 1$ without symmetries. But for three dimensions this method fails.
It is hard to guess whether there is a unique classical solution of (2) in 3 dimensions. The situation is similar to the situation in the field of Navier-Stokes equations:

In 3 dimensions one has existence but not uniqueness for weak solutions, uniqueness but not existence for classical solutions; in lower dimensions existence and uniqueness of classical solutions.

What about higher dimensions? Nothing is known for Navier-Stokes and nothing is known for Vlasov in the plasmaphysical case. But - here is the only essential difference between $\gamma = +1$ and $\gamma = -1$ - Horst has shown, that there is no global solution in 4 dimensions for the stellardynamic case. Since that result is at least mathematically surprising and since the proof is simple, I state it as

Theorem 5: In 4 dimensions for $\gamma > 0$ there are initial values $\overset{o}{f}$ (positive, continuously differentiable with compact support), such that the corresponding initial value problem for (L) has no global solution.

Remark: 4 dimensions mean 4 space and 4 velocity variables. G has to be then

$$G(x,y) = -\gamma \frac{x-y}{\|x-y\|^4}, \quad x,y \in \mathbb{R}^4.$$

Proof: We consider the moment of inertia for a solution of (2) (for times t, when it exists).

$$(24) \qquad mi(t) := \int \|x\|^2 f(t,x,v)\,dxdv$$

With V defined like in (17) and with f instead of f_δ and

$$U(x-y) = \frac{\gamma}{2} \frac{1}{\|x-y\|^2}$$

instead of U_δ (V is the potential energy) one gets by straightforward calculation, that mi is twice continuously differentiable and

$$mi''(t) = 2(E(t) + V(t)) = 2(E(0) + V(0))$$

(here the dimension 4 comes in; in 3 dimensions it would just be $mi''(t) = 2E(t) + V(t)$ and we could not use the energy conservation).

Now choose a $\overset{\circ}{f}$ such that $E(0) + V(0) < 0$. This is possible only in the stellardynamic case, where $\gamma > 0$, since E(0) is positive but linear in $\overset{\circ}{f}$ and V(0) is negative but quadratic in $\overset{\circ}{f}$. Therefore mi"(t) is a negative constant. mi itself is positive, since $\overset{\circ}{f}$ and therefore f is positive. This is possible only in certain finite time intervals. So the solution can only exist in a finite time interval (which depends only on $\overset{\circ}{f}$). That was to be shown.

4. The plasmaphysical case with selfconsistent magnetic field

If the particles considered are electrons, they do not only interact by means of the electric field but also by the magnetic field generated by themselves. Therefore instead of looking only at the electric field

$$E(t,x) = \int_{\mathbb{R}^3} G(x,y)f(t,y,v)\,dydv = \int_{\mathbb{R}^3} G(x,y)\rho(t,y)\,dy$$

in (2), which is a solution of $\operatorname{div} E = -4\pi\gamma\rho$, we have to take into account the full Maxwell equations. In order to be in accordance with a familiar form of these equations, we slightly change the notation.

Instead of (1), (2) the system of equations we have to consider now is the following system for the function f, the electric field E and the magnetic field B:

(2') $\frac{\partial f}{\partial t} + \langle v, \operatorname{grad}_x f\rangle - \langle E[f] + \frac{1}{c}\, v \times B[f], \operatorname{grad}_v f\rangle = 0$

(1a) $\operatorname{div} E = 4\pi(n_e + n_i)$

(1b) $\operatorname{div} B = 0$

(1c) $\operatorname{rot} E = -\frac{1}{c}\frac{\partial B}{\partial t}$

(1d) $\operatorname{rot} B = \frac{1}{c}\frac{\partial E}{\partial t} + \frac{4\pi}{c} j$

where

(25) $n_e(t,x) := -\int_{\mathbb{R}^3} f(t,x,v)\,dv, \quad j(t,x) := -\int_{\mathbb{R}^3} vf(t,x,v)\,dv$

and $n_i = n_i(x)$ is a given spatial density of a fixed ion background.

Remarks:

(a) If $B = 0$ and $n_i = 0$, then (2') is essentially (2), provided we add to (1a) the boundary condition that E has to vanish at infinity. Then (1a) has the solution

$$E(t,x) = \int G(x,y)n_e(t,y)\,dy$$

with $\gamma = -1$.

(b) In (2') we write E[f] and B[f] in order to make clear that the fields depend on f: n_e and j are moments of f, therefore a solution of the system (1a) - (1d) depends on f.

(c) We introduced the function n_i representing the spatial density of a ion gas, while f is the μ-space density for the electrons. We assume that the ions form a fixed background. Otherwise we would have to consider two distribution functions f_e and f_i and two equations of the form (2'), which differ only by a constant and a sign in front of the third term. This would not cause principal difficulties but increase the formal complexity of the system.
We mention that the introduction of a n_i and even of an external magnetic field, which does not depend on f, into equations (1) - (2) of the preceding lectures would not have generated essential troubles and could have been handled easily.

In order not to get lost in the complexity of the problem we are looking for the simplest geometry in which the typical features of the problem are present. If f would depend only on one space and one velocity variable no magnetic field would occur. Therefore we look for solutions which depend only on one space and on two velocity variables:

$$f(t,\cdot) : (x_1,v_1,v_2) \longrightarrow f(t,x_1,v_1,v_2) \geq 0 \quad \text{for all } P := (x_1,v_1,v_2) \in \mathbb{R}^3.$$

For such a function n_e and j defined by (25) do not exist. Therefore we change these definitions into

$$(25') \quad n_e(t,x_1) = -\int_{\mathbb{R}^2} f(t,x_1,v)dv, \quad j(t,x) = -\int_{\mathbb{R}^2} vf(t,x_1,v)dv$$

with $v = (v_1,v_2)$.

Furthermore, we add some boundary conditions for E and B, which also will depend only on x_1:

$$(26a) \quad \lim_{x_1 \to -\infty} E_1(t,x_1) = -\lim_{x_1 \to +\infty} E_1(t,x_1), \quad t \in [0,T]$$

$$(26b) \quad E_2(0,x_1) = \overset{0}{E}_2(x_1), \quad E_3(0,x_1) = 0 \quad \text{for all } x_1 \in \mathbb{R}$$

$$(26c) \quad B_1(0,x_1) = B_2(0,x_1) = 0, \quad B_3(0,x_1) = \overset{0}{B}_3(x_1) \quad \text{for all } x_1 \in \mathbb{R}.$$

It follows that $B_1(t,x_1) = B_2(t,x_1) = E_3(t,x_1) = 0$ everywhere and instead of (1a) - (1d) the simpler system

$$(1a') \qquad \frac{\partial E_1}{\partial x_1} = 4\pi (n_e+n_i)$$

$$(1c') \qquad \frac{\partial E_2}{\partial x_1} + \frac{1}{c}\frac{\partial B_3}{\partial t} = 0$$

$$(1d') \qquad \frac{\partial E_1}{\partial t} = -4\pi j_1, \quad \frac{\partial E_2}{\partial t} + c\frac{\partial B_3}{\partial x_1} = -4\pi j_2.$$

remains to be considered.

(2') becomes

$$(2") \qquad \frac{\partial f}{\partial t} + v_1 \frac{\partial f}{\partial x_1} - (E_1[f] + \frac{1}{c} v_2 B_3[f]) \frac{\partial f}{\partial v_1} - (E_2[f] - \frac{1}{c} v_1 B_3[f]) \frac{\partial f}{\partial v_2} = 0$$

The initial value problem for the system (2'), (1a'), (1c') and (1d') is unmodified in the sense of lecture 1 but lower dimensional. I suggest that it has a unique classical solution. But this is not yet proved. Only for a "mollified" problem existence and uniqueness is assured and I will sketch that result (see [ ]).

Instead of defining n_e and j as in (25') we use a smoothed f_δ instead of f and define

$$(25") \qquad n_e^\delta(t,x_1) = - \int_{\mathbb{R}^3} \omega_\delta(x_1-\xi)\, f(t,\xi,v)\, d\xi\, dv$$

$$j^\delta(t,x_1) = - \int_{\mathbb{R}^3} v\omega_\delta(x_1-\xi)\, f(t,\xi,y)\, d\xi\, dv.$$

Now our problem is completely stated by the equations (1a'), (1c'), (1d'), (2") and (25"), where into the equations (1') n_e^δ, j^δ must be inserted instead of n_e, j. We have to add the initial and boundary conditions (26) for B, E and f.

We first solve the equations (1a'), (1c') and (1d') for given $f(t,\cdot)$. Let μ_t be again the measure with the density $f(t,\cdot)$. We have to be a little bit more careful, as μ_t has to be such that

$$j^{\delta}(t,x_1) = \int_{\mathbb{R}^3} v\omega_{\delta}(x_1-\xi)\, d\mu_t(\xi,v)$$

exists. But the function $(\xi,v) \longrightarrow v\omega_{\delta}(x_1-\xi)$, x_1 fixed, is not bounded. That causes some trouble, in contrast to the situation of lecture 1.

We will therefore assume, that the supports of μ_t, $t \in [0,T]$ are uniformly bounded with respect to t; i.e. there exists a ball $\kappa_R = \{P \in \mathbb{R}^3 / \|P\| \leq R\}$ such that

$$\text{supp}\ \mu_t \in \kappa_R \quad \text{for } t \in [0,T].$$

Now (1a') with the boundary condition (26a) has the solution

$$E_1(t,x_1) = 2\pi \int_{-\infty}^{+\infty} \text{sign}(x_1-y)[n_e^{\delta}(t,y) + n_i(y)]dy$$

$$= 2\pi \int_{\mathbb{R}^3} \left(\int_{-\infty}^{+\infty} \text{sign}(x_1-y)\omega_{\delta}(y-z)\, dy\right) d\mu_t(z,v)$$

$$+ 2\pi \int_{-\infty}^{+\infty} \text{sign}(x_1-y) n_i(y)\, dy .$$

It is easy to check, that E_1 is continuous and bounded in $[0,T] \times \mathbb{R}$, is globally Lipschitz continuous with respect to x with a Lipschitz constant independent of t and $\mu_{\cdot}$. Furthermore, by doing the same small calculations as in lecture 1 one gets

$$|E_1[\mu.](t,x) - E_1[\nu.](t,x)| \leq K_1 \rho(\mu_t,\nu_t).$$

For arbitrary $\mu.$ the function E_1 given by (27) will in general not satisfy the first equation in (1d'). We forget that equation for a moment; it will turn out at the end that, if $\mu.$ is a solution of our problem, this equation is automatically satisfied.
Now we have to consider the equations

$$\frac{\partial E_2}{\partial x_1} + \frac{1}{c}\frac{\partial B_3}{\partial t} = 0, \quad \frac{\partial E_2}{\partial t} + c\frac{\partial B_3}{\partial x} = -4\pi j_2^{\delta}$$

with the initial condition $E_2(0,x_1) = \overset{\circ}{E}_2(x_1)$, $B_3(0,x_1) = \overset{\circ}{B}_3(x_1)$.

To solve them we look for a function $A\colon [0,T] \times \mathbb{R} \longrightarrow \mathbb{R}$, so that

$$(28) \qquad B_3 = \frac{\partial A}{\partial x_1}, \quad E_2 = -\frac{1}{c}\frac{\partial A}{\partial t}$$

is a solution. Inserting this "Ansatz" into the equations for E_2, B_3 we get

$$(29) \qquad \frac{\partial^2 A}{\partial x_1^2} - \frac{1}{c^2}\frac{\partial^2 A}{\partial t^2} = -\frac{4\pi}{c} j_2^\delta$$

i.e. the inhomogenous wave equation. Its solution is

$$(30) \qquad \begin{aligned} A(t,x_1) &= -2\pi \int_0^t \Big(\int_{x_1-c(t-\tau)}^{x_1+c(t-\tau)} j_2^\delta(\tau,y)\, dy\Big)\, d\tau - \frac{1}{2}\int_{x_1-ct}^{x_1+ct} \overset{o}{E}_2(y)\, dy \\ &+ \frac{1}{2}\Big(\int_{-\infty}^{x_1+ct} \overset{o}{B}_3(y)\, dy + \int_{-\infty}^{x_1-ct} \overset{o}{B}_3(y)\, dy\Big) . \end{aligned}$$

E_2 and B_3 given by (28),(30) satisfy similar conditions as E_1: They are globally Lipschitz continuous uniformly with respect to t and μ. and there are constants K_2, K_3, M_2, M_3, such that

$$(31) \qquad \begin{aligned} |E_2[\mu.](t,x_1) - E_2[\nu.](t,x_1)| &\le K_2\rho(\mu_t,\nu_t) + M_2 \int_0^t \rho(\mu_\tau,\nu_\tau)\, d\tau \\ |B_3[\mu.](t,x_1) - E_3[\nu.](t,x_1)| &\le K_3\rho(\mu_t,\nu_t) + M_3 \int_0^t \rho(\mu_\tau,\nu_\tau)\, d\tau \quad . \end{aligned}$$

However, these constants K_2, K_3, M_2, M_3 depend on the bound of the supports of μ_t, i.e. on R.

The vectorfield $V[\mu.]$, defined by

$$V[\mu.](t,x_1,v) := \begin{pmatrix} v_1 \\ -E_1[\mu.] - \frac{1}{c} v_2 B_3[\mu.] \\ -E_2[\mu.] + \frac{1}{c} v_1 B_3[\mu.] \end{pmatrix} .$$

satisfies condition (I) of lecture 1; condition (II) has to be generalized to

$$(\mathrm{II}') \qquad \int_{\mathbb{R}^k} \| V[\mu.](t,P) - V[\nu.](t,P) \| \, d\mu_t(P) \le$$

$$\le K\rho(\mu_t,\nu_t) + M \int_0^t \rho(\mu_\tau,\nu_\tau)\, d\tau .$$

There $k=3$; the additional term $M\int_0^t \rho(\mu_\tau,\nu_\tau)\,d\tau$ causes no trouble in any proof. What is really unpleasant is the fact, that (I) and (II') are only satisfied, if μ_t,ν_t, $t\in[0,T]$ have uniformly bounded support. We need an a priori estimate: If $\mu.$ is in $[0,T]$ a solution of the initial value problem with $\overset{\circ}{\mu}$ having bounded support, then μ_t, $t\in[0,T]$ has uniformly bounded support.

In order to prove this we consider again the kinetic energy

$$(16')\qquad W(t) := \int_{\mathbb{R}^3} \|v\|^2\, d\mu_t(x_1,v)$$

(We use W instead of E in order to avoid confusion with the electric field). Using the fact, that $\mu.$ is a solution of (2") we get

$$\dot W(t) = -2\int_{\mathbb{R}^3} (v_1E_1(t,x_1) + v_2E_2(t,x_1))\,d\mu_t(x_1,v)\,.$$

From (27),(28) and (30), we get

$$W(t) \le W(0) + 2\alpha\left(t + \int_0^t W(\tau)\,d\tau\right) + 4\pi c_\delta t\cdot$$

$$\cdot\int_0^t\left(\int_{\mathbb{R}^3} |v_2|^2\, d\mu_\tau(x_1,v)\right)d\tau$$

(with $\alpha = c+4\pi$) and finally

$0\le W(t)\le C_1+C_2\int_0^t W(\tau)\,d\tau$, where C_1,C_2 are dependent on $W(0)$, T and δ.

An application of Gronwall's lemma yields $0\le W(t)\le A$ for $t\in[0,T]$, where A depends only on $W(0)$, δ and T.

As $\int_{\mathbb{R}^3} |v_i|\,d\mu_t(x_1,v)\le 1+W(t)$, $i=1,2$, we get a uniform bound for these integrals and therefore also for E_1,E_2 and B_3. Hence

$$\|V[\mu.](t,P)\|\le L_1+L_2\|P\| \quad \text{for } t\in[0,T],\ P\in\mathbb{R}^3$$

with constants depending only on $\overset{\circ}{\mu},\delta$ and T. It follows that

$$\|T_{t,0}[\mu.]Q\|\le(\|Q\|+L_1T)e^{L_2T},\quad t\in[0,T],\ Q\in\mathbb{R}^3$$

Choosing $R := (r+L_1T)e^{L_2T}$, we see that for $Q\in\operatorname{supp}\overset{\circ}{\mu}$ the point

$T_{t,0}[\mu.]Q$ is contained in κ_R and therefore

$$\operatorname{supp} \mu_t = \operatorname{supp}(\overset{\circ}{\mu} \circ T_{0,t}[\mu.]) \subset \kappa_R.$$

This gives the desired a priori estimate.

The remaining part of an uniqueness and existence proof is simple: One first shows a local (with respect to time) theorem bý one of the standard methods and continues the solution by using the a priori estimate. The details can be found in [ ]. This result is

Theorem 6: For $\overset{\circ}{f} \in L^1(\mathbb{R}^3)$ with compact support the corresponding initial value problem given by (2"), (1a'), (1c'), (1d'), (25") together with the initial-boundary condition (26) has in any intervall $[0,T]$ a unique (weak) solution. The solution is classical if the initial data $\overset{\circ}{f}$, $\overset{\circ}{E}_2$ and $\overset{\circ}{B}_3$ are continuously differentiable.

Remark:

The following questions arise but are not answered till now.

(1) Could a similar theorem be proved without reducing the dimension of the problem, i.e. for the sixdimensional "mollified" case? Since no essential use of the dimension is made and the advantage consists only in getting solutions of the Maxwell equations, which are easy to handle, there should be a good chance for a positive answer.

(2) Is it possible to take the limit $\delta \longrightarrow 0$ to get solutions for the unmodified equations? It should be by far easier to get existence but not uniqueness of weak solutions. (Inspite of not having shown, that a δ-independent bound for the kinetic energy W exists) than to prove something about classical solutions.

Theorem 6 might be considered only as a first step in solving a problem, which is of high interest for applications in plasma physics.

5. Numerical Methods, in particular Simulation Procedures

There exist a lot of numerical methods in order to solve the Vlasov equation approximately. Difference schemes, Fourier expansions with respect to x, Hermite polynomials are used as well as the so-called waterbag model, which uses the fact that if $\overset{\circ}{f}$ is a stepfunction, then the solution $f(t,\cdot)$ is of the same kind. As far as I know, for none of these methods convergence is proved. Besides these methods plasmaphysicists very often use so-called simulation methods, especially the "particle-in-cell"-method (PIC), developed for the Vlasov equation by R. Morse. Since the convergence of this method can be proved by the methods developed in lecture 1, we will describe it in more details.
The main idea of a simulation in the kinetic theory is simply as follows: A real gas consists of say 10^{23} particles. The Vlasov equation describes a gas of infinitely many particles (in the sense given in lecture 1). Let's try to create a gas of about 10^4 particles, whose motion is governed by the rules of classical mechanics such that this gas behaves "as similar as possible" to the infinite particle system.
In order to make this more precise, we reformulate the idea:

Let N be the number of particles in a real system, $\overset{\circ}{\omega}_N$ the initial state and $\overset{\circ}{\mu} = \int \overset{\circ}{f}\, dP$ a good approximation for $\delta_{\overset{\circ}{\omega}_N}$.

Let n be the number of particles in the "simulation gas". Now let us try to find an initial state $\overset{\circ}{\omega}_n$ for these particles, such that the distance

$$\rho(\delta_{\omega_n(t)}, \mu_t)$$

of the state of the simulated n-particle system at time t to the solution of the Vlasov equation at the same time is as small as possible.

It should be clear that the only possibility to be successful is in constructing $\overset{\circ}{\omega}_n$ for given $\overset{\circ}{f}$. Then $\delta_{\omega_n(t)}$ is given by solving the Newtonian system (3), which can be done at least numerically in a satisfying manner, if n is not too large. But it is again necessary to smooth the interaction forces or, equivalently, to "smear out" the particles. This is done by PIC - the particles are smeared out over cells.

We want to get some information about how the complexity of the method increases if we increase the dimension k; $k=3$ is the normal case of 3 space and 3 velocity variables. The interaction force is then given by

$$G(x,y) = -\gamma \frac{x-y}{\|x-y\|^k}, \quad k=1,2,3.$$

The steps of PIC are:

Step 1: Choose $\overset{o}{\omega}_n$ such that $\rho(\delta_{\overset{o}{\omega}_n}, \overset{o}{\mu})$ is as small as possible ("as small as possible" means: as small as you can make it):

$$\overset{o}{\omega}_n = \{(\overset{o}{x}_1,\overset{o}{v}_1),\dots,(\overset{o}{x}_n,\overset{o}{v}_n)\}.$$

Step 2: Instead of ω_δ given in lecture 1 we choose a less smooth but simpler mollifier

$$\omega_\delta(x) = \begin{cases} \dfrac{1}{\gamma_k \delta^k} & \text{for } \|x\| \le \delta \\ 0 & \text{otherwise} \end{cases},$$

where γ_k is the volume of the k-dimensional unit ball, $\gamma_1 = 2$, $\gamma_2 = \pi$, $\gamma_3 = \frac{4\pi}{3}$.

We use this ω_δ in order to smear out the particles over balls of radius δ in the x-space (i.e. over "cells"); therefore we substitute the spatial distribution

$$\frac{1}{n}\sum_{j=1}^{n} \delta_{\overset{o}{x}_j}$$

of the n-particle system by a distribution with the spatial density

$$(32) \quad \overset{o}{\rho}_n(x) := \frac{1}{n}\sum_{j=1}^{n} \omega_\delta(x-\overset{o}{x}_j) = \int_{\mathbb{R}^k} \omega_\delta(x-z)\,d\delta_{\overset{o}{\omega}_n}(z,v)$$

Then

$$\int_{\mathbb{R}^k} G(x,y)\overset{o}{\rho}_n(y)\,dy = \int_{\mathbb{R}^k} G(x,y)\Big(\int_{\mathbb{R}^k} \omega_\delta(y-z)\,d\delta_{\overset{o}{\omega}_n}(z,v)\Big)dy$$

$$= \int_{\mathbb{R}^k} \left(\int_{\mathbb{R}^k} G(x,y)\,\omega_\delta(y-z)\,dy \right) d\delta_{\overset{o}{\omega}_n}(z,v)$$

$$= \int_{\mathbb{R}^k} G_\delta(x,z)\,d\delta_{\overset{o}{\omega}_n}(z,v)$$

so that this procedure is equivalent to our former modification, where we substitute G by G_δ.

Remark: Our procedure consists in smearing out a particle over a δ-ball centered at the particle. The normal PIC uses fixed cells not depending on the positions of the particles, but adds afterwards a socalled area weighting, which gives at least for k=1 exactly the same values of $\overset{o}{\rho}_n$. For $k > 1$ the differences are not important and can be avoided by using another ω_δ.

Step 3: Calculate

$$\overset{o}{E}_n(x) = E_n(0,x) = \int G(x,y)\,\overset{o}{\rho}_n(y)\,dy$$

for $x = \overset{o}{x}_j$, j=1,...,n. Since $\overset{o}{\rho}_n$ is a step function, the integration can be done explicitly; one only has to evaluate n values of an explicit function.

Step 4: Propagate the particles for a time step $\Delta t > 0$ by

$$\hat{v}_i(\Delta t) = \overset{o}{v}_i + \Delta t\, E_n(0,\overset{o}{x}_i)$$
$$\hat{x}_i(\Delta t) = \overset{o}{x}_i + \Delta t\, v_i(\Delta t) \tag{33}$$

This provides the first step of a numerical integration of the characteristic equations

$$\dot{x} = v$$
$$\dot{v} = E[\delta_{\omega_n(\cdot)}](t,x).$$

$(\hat{x}_j(\Delta t), \hat{x}_j(\Delta t))$ is an approximation for $T_{\Delta t,0}[\delta_{\omega_n(\cdot)}]\overset{o}{P}_i$.

Choosing the symmetric difference scheme (33) has some advantages. The mapping

$$(\overset{o}{x}_i, \overset{o}{v}_i) \longrightarrow (\hat{x}_i(\Delta t), \hat{v}(\Delta t)) = \hat{P}_i(\Delta t)$$

has some of the properties $T_{\Delta t,0}[\mu_\cdot]$ has: It is measure preserving and bijective.

Step 5: Repeat the process with $(\hat{P}_1(\Delta t),\ldots,\hat{P}_n(\Delta t))$ instead of $(\overset{o}{P}_1,\ldots,\overset{o}{P}_N)$ in order to get $\rho_n(\Delta t,x)$, $E_n(\Delta t,x)$ and finally $\hat{P}_1(2\Delta t),\ldots,\hat{P}_n(2\Delta t))$ and so on.

Remark: As the numerical method is uniformly convergent for $\Delta t \to 0$, we have $\|P_i(m\Delta t)-\hat{P}_i(m\Delta t)\| \leq \varepsilon$ for $0 \leq m\Delta t \leq T$, if Δt is small enough. Hence

$$\rho(\delta_{\omega_n}(t),\delta_{\hat{\omega}_n}(t)) = \sup_{\varphi\in D}\left| \int \varphi\, d\delta_{\omega_n}(t) - \int \varphi\delta_{\hat{\omega}_n}(t)\right|$$

$$\leq \sup_{\varphi\in D} \frac{1}{n}\sum_{j=1}^{n} |\varphi(P_i(t)) - \varphi(\hat{P}_i(t))|$$

$$\leq \max_{i=1,\ldots,n} \|P_i(t) - \hat{P}_i(t)\| < \varepsilon$$

for $t = m\Delta t$, Δt small enough.

According to theorem 1

$$\rho(\mu_t,\delta_{\omega_n}(t)) \leq e^{Ct}\rho(\overset{o}{\mu},\delta_{\overset{o}{\omega}_n});$$

therefore

$$\rho(\mu_t,\delta_{\hat{\omega}_n}(t)) \leq \rho(\mu_t,\delta_{\omega_n}(t)) + \rho(\delta_{\omega_n}(t),\delta_{\hat{\omega}_n}(t)) \leq \varepsilon + e^{Ct}\rho(\overset{o}{\mu},\delta_{\overset{o}{\omega}_n}).$$

Consequently, PIC converges in the sense that for given $\varepsilon > 0$, $T > 0$ one can choose Δt small enough and n large enough such that

$$\rho(\mu_t,\delta_{\hat{\omega}_n}(t)) < \varepsilon \qquad \text{for } t = m\Delta t,\ 0 \leq m\Delta t \leq T.$$

This entails that the numerically calculated electric field E_n approximates the real electric field E uniformly, i.e.

$$\|E_n(m\Delta t,\cdot) - E(m\Delta t,\cdot)\|_0 < \varepsilon \qquad \text{for } 0 \leq m\Delta t \leq T,\ \Delta t \text{ sufficiently small.}$$

Looking a bit more carefully, how the constant $C = K+L$ in the preceding estimate depends on δ and the dimension k, one gets

$$\rho(\mu_t,\delta_{\omega_n}(t)) \leq e^{C(k,\delta)t}\rho(\overset{o}{\mu},\delta_{\overset{o}{\omega}_n})$$

with

$$C(k,\delta) = \sqrt{2}\left(1 + \frac{\sqrt{k}}{\delta^k}\left(1 - \frac{\sqrt{2}}{\delta^{k-1}}\right)\right).$$

If δ is chosen near to 1 (i.e. in physical dimensions: approximately equal to the socalled Debye length), then $C(k,\delta)$ is of order $\sqrt{k}$.

The only thing which remains to be done and which is of big practical importance is to find an appropriate $\overset{\circ}{\omega}_n$ for given $\overset{\circ}{\mu}$. The problem of finding the optimal $\overset{\circ}{\omega}_n$ with respect to all n-particle systems has not even been attacked.
So we will discuss a method to construct a "good" (not a best) $\overset{\circ}{\omega}_n$ for given $\overset{\circ}{\mu} = \int \overset{\circ}{f}\, dP$. For the special case $\overset{\circ}{f} = \chi_E$, where E is the unit cube in $\mathbb{R}^{2k}$, $E = \{P \in \mathbb{R}^{2k} \mid 0 \le p_i < 1,\ i=1,\dots,2k\}$ and χ_E is the characteristic function of E, there are some results available. Therefore we reduce the general case to the special one.

We assume, that there exists a convex domain B in $\mathbb{R}^{2k}$ such that $\overset{\circ}{f}(P) > 0$ for $P \in B$, $\overset{\circ}{f}(P) = 0$ for $P \notin B$. Then one is able to construct a mapping $T : B \to E$, which has the following properties:
T is differentiable and bijective, and its Jacobian is

$$J_T(P) = \overset{\circ}{f}(P),\quad P \in B.$$

Then $\overset{\circ}{\mu} = \mu_E \circ T$, where $\mu_E = \int \chi_E dP$ is the uniform distribution in E:

$$\overset{\circ}{\mu}(M) = \int_B \chi_M \overset{\circ}{f}\, dP = \int_B \chi_M J_T dP = \int_E \chi_M \circ T^{-1} dP = \int_E \chi_{T(M)} dP = \mu_E(T(M))$$

for arbitrary M.
The construction of T is given in a paper by Hlawka and Mück in 1972. It is simple when $\overset{\circ}{f}$ factorises, i.e.

$$\overset{\circ}{f}(P) = \overset{\circ}{f}_1(p_1) \dots \overset{\circ}{f}_{2k}(p_{2k}) \qquad (p_i = x_i \text{ and } p_{k+i} = v_i \text{ for } i=1,\dots,k).$$

Then

$$T_i(P) := T_i(p_i) = \int_{-\infty}^{p_i} \overset{\circ}{f}_i(\xi)d\xi,\ i=1,\dots,2k,\ P \in B.$$

If now $\tilde{\omega}_n = \{\tilde{P}_1,\dots,\tilde{P}_n\}$ is a good approximation for μ_E, then $\overset{\circ}{\omega}_n := \{T^{-1}\tilde{P}_1,\dots,T^{-1}\tilde{P}_n\}$ is a good approximation for $\overset{\circ}{\mu}$: If T^{-1} is Lipschitz continuous with a Lipschitz constant λ , then

$$\rho(\delta_{\overset{\circ}{\omega}_n}, \overset{\circ}{\mu}) \le \lambda \rho(\delta_{\tilde{\omega}_n}, \mu_E).$$

T^{-1} is relatively easy to calculate in the special case mentioned above. The discrepancy, defined in (6) is also a metric which measures the distance of $\delta_{\overset{o}{\omega}_n}$ to $\overset{o}{\mu}$. For the purposes discussed here it is easier to handle. For example, for a factorising $\overset{o}{f}$, a set $R \in \mathfrak{R}$ is transformed by T into a set of the same kind whose corner Q is in E. It follows that for these kind of initial conditions even the quality

$$D(\delta_{\overset{o}{\omega}_n}, \overset{o}{\mu}) = D(\delta_{\tilde{\omega}_n}, \mu_E)$$

holds.

Therefore, it remains to construct $\delta_{\tilde{\omega}_n}$. We have to point out, that this problem cannot be solved in a nice way by using a random generator: The only property $\tilde{\omega}_n$ has to have is, that the discrepancy $D(\delta_{\tilde{\omega}_n}, \mu_E)$ is small, therefore things like correlations are irrelevant. $\tilde{\omega}_n$ might be distributed very regularly - this is what physicists call a "quiet start".

There is a method proposed by Niedereiter how to calculate $D(\delta_{\tilde{\omega}_n}, \mu_E)$ without taking the l.u.b. over all $R \in \mathfrak{R}$; a finite number of property selected R is enough.

Just to give you an example how real calculations are done, we consider the one dimensional case k=1. For n we choose only the Fibonacci numbers $n=\alpha_k$, $k \in \mathbb{N}$ with $\alpha_o = \alpha_1 = 1$, $\alpha_{k+1} = \alpha_k + \alpha_{k-1}$.

Then we choose $\{\tilde{P}_1, \ldots, \tilde{P}_n\}$ with $\tilde{P}_i = (x_i, v_i)$ in the following way:

$$x_i = \frac{2i-1}{2\alpha_k}, \quad v_i = \left[\frac{2(i-1)\alpha_{k-1}+1}{2\alpha_k}\right], \quad i=1,\ldots,\alpha_k=n,$$

where $[\xi]$ denotes the fractional part of the positive real number ξ.

One obtains the following values for $D(\delta_{\tilde{\omega}_n}, \mu_E)$

k	$\alpha_k=n$	$D(\delta_{\tilde{\omega}_n}, \mu_E)$
12	144	$1,5 \cdot 10^{-2}$
14	377	$7,6 \cdot 10^{-3}$
16	987	$3,3 \cdot 10^{-3}$
18	2584	$1,4 \cdot 10^{-3}$
20	6765	$5,6 \cdot 10^{-4}$

For 6765 points, generated by the IBM-random-generator, the discrepancy is $1,4\cdot 10^{-2}$; that means, that you need only 144 points with the construction given above to obtain a similar result. The computing time however differs by a factor 47.

In the table given above one realizes that the convergence of $D(\delta_{\tilde{\omega}_n}, \mu_E) = D_n$ to zero is rather slow. In fact it has been shown by numbertheoretists, that there exists a constant C_k, depending only on the dimension, such that

$$D_n \geq \frac{C_k}{n} (\ln n)^{\frac{k-1}{2}}$$

So, one cannot expect a very quick convergence of the method. Nevertheless, it is better than most of the other methods used in praxis - and its convergence is proved!

6. Stationary Solutions

Not very much is known about stationary (i.e. time independent) solutions. The only exeption is the one-dimensional case, where all solutions can be explicitly constructed. These solutions, discovered in 1957, are the socalled "Bernstein-Green-Kruskal"-modes (BGK-modes).

The problem is to construct all solutions of the system

$$\text{(34a)} \qquad v \frac{\partial f_-}{\partial x} + E(x) \frac{\partial f_-}{\partial v} = 0$$

$$\text{(34b)} \qquad v \frac{\partial f_+}{\partial x} - E(x) \frac{\partial f_+}{\partial v} = 0$$

$$\text{(35)} \qquad c_- \int_{-\infty}^{+\infty} f_-(x,v)\,dx - c_+ \int_{-\infty}^{+\infty} f_+(x,v)\,dv = E'(x),$$

where c_- und c_+ are some positive constants. f_-, f_+ are the distribution functions of the electrons and the ions respectively. We consider here a two-component gas, because in this case the solutions become even simpler.

By solving (35) together with some boundary values for E and substituting the solution into (34) one realizes that the problem is again nonlinear.

However, the main idea is to make the problem linear by prescribing appropriate data of f. If we look for solutions of (34), (35) with

(36) $$\rho(x) := c_- \int_{\infty}^{\infty} f_-(x,v)\,dv - c_+ \int_{-\infty}^{+\infty} f_+(x,v)\,dx$$

given, then E'(x) is known from the data, E is - up to a boundary condition - a known function and we have to solve the linear equation (34) with the rather unusual condition (36).

For a given E, let us consider (34a). As f_- depends only on two variables x and v, we know that a function f_- is a solution of (34a) if and only if it depends only on the Hamiltonian

(37) $$H(x,v) = v^2 - 2U(x)$$

where U is an arbitrary integral of E

(38a) $$f_-(x,v) = \phi_-(v^2 - 2U(x)).$$

Similarly f_+ has to be of the form

(38b) $$f_+(x,v) = \phi_+(v^2 + 2U(x)).$$

The figure on the following page gives you a typical picture for the level curves of f_+, f_- for a given U. To different branches of such a level curve one might assign different values of the functions $f_\pm$. These level curves for f_+ degenerate to a single point at x_1, where U has a local minimum, if we choose $c = 2U(x_1)$. For $c < 2U(x_1)$ the level curve is an empty set.
The level curves have an intersection with the x-axis, if $\frac{c}{2}$ is contained in the range of U; at these intersection points, the level curves are orthogonal to the x-axis.

The level curves for f_- behave somewhat antisymmetric. They degenerate at maximum points x_o of U, do not exist for $c < -2U(x_o)$ and intersect with the x-axis, if $\frac{c}{2}$ is contained in the range of -U.

We restrict our considerations for a moment to an interval of strict monotonicity of U, say $[x_o, x_1]$ and choose an arbitrary ξ in (x_o, x_1).

Every point (x,v) in the strip $[\xi, x_1] \times \mathbb{R}$ lies on a level curve of $f_\pm$, which passes $x = \xi$. This is not true for the level curves of f_+.

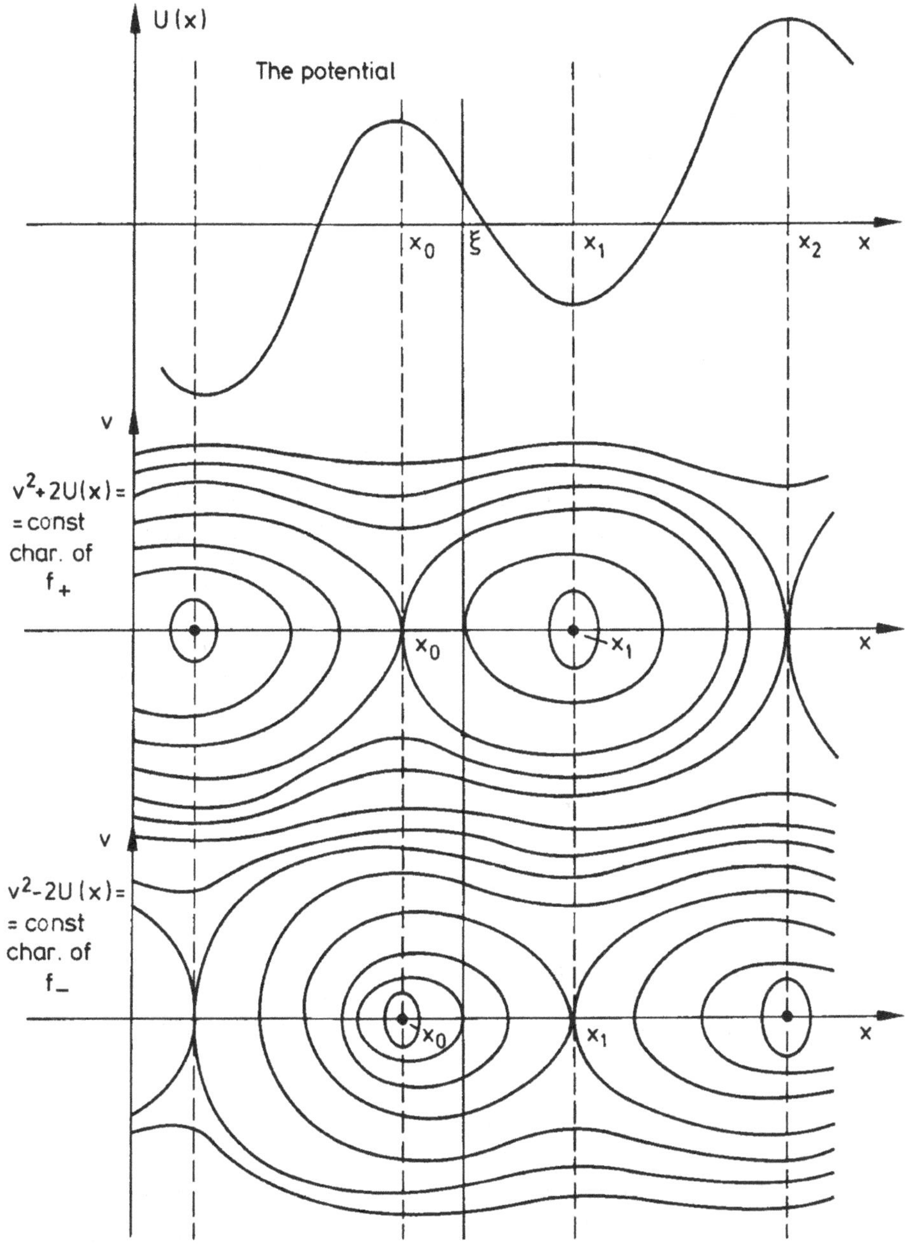
U(x)
The potential
x_0
ξ
x_1
x_2
x
v
v²+2U(x)=
= const
char. of
f_+
x_0
x_1
x
v
v²-2U(x)=
= const
char. of
f_-
x_0
x_1
x

The level curve passing $(\xi,0)$ is given by

$$v^2 + 2(U(x)-U(\xi)) = 0.$$

Therefore points $(x,v) \in [\xi,x_1] \times \mathbb{R}$ with

$$|v| < \sqrt{2(U(\xi)-U(x))}$$

are not reached by curves passing $(\xi,0)$.

If, in addition to ρ according to (36), we prescribe the boundary values

$$(39) \qquad f_+(\xi,v) = \varphi_+(v), \quad f_-(\xi,v) = \varphi_-(v),$$

then f_- is completely determined in $[\xi,x_1] \times \mathbb{R}$; f_+ is determined only for (x,v) with $\xi \le x \le x_1$, $|v| \ge \sqrt{2(U(\xi)-U(x))}$.

Therefore $\int_{-\infty}^{+\infty} f_-(x,v)dv$ is determined by φ_-, $\int_{|v| \ge \sqrt{2(U(\xi)-U(x))}} f_+(x,v)dv$ by φ_+.

In order to satisfy (36), we have to solve the equation

$$\int_{-\sqrt{2(U(\xi)-U(x))}}^{+\sqrt{2(U(\xi)-U(x))}} f_+(x,v)dv = -\frac{\rho(x)}{c_+} + \frac{c_-}{c_+}\int_{-\infty}^{+\infty} f_-(x,v)dv - \int_{|v| \ge \sqrt{2(U(\xi)-U(x))}} f_+(x,v)dv =: h(x)$$

h is known in $[\xi,x_1]$; using (38b), we get

$$2\int_0^{\sqrt{2(U(\xi)-U(x))}} \phi_+(v^2+2U(x))dv = h(x) \; .$$

With $v^2+2U(x) = t$, we find

$$\int_{2U(x)}^{2U(\xi)} \phi_+(t) \frac{dt}{\sqrt{t-2U(x)}} = h(x).$$

U is strictly monotone in x; so 2U has an inverse function W and we obtain

$$\int_{y}^{2U(\xi)} \phi_+(t) \frac{dt}{\sqrt{t-y}} = h(W(y)).$$

This is Abel's integral equation, which has the unique solution

$$(40) \qquad \phi_+(z) = \frac{1}{\pi}\frac{d}{dz}\int_z^{2U(\xi)} \frac{h(W(t))}{\sqrt{z-t}}, \quad 2U(x) \le z \le 2U(\xi).$$

We do not want to work out all the details. However, we note that $f_+(x,v) = \phi_+(v^2+2U(x))$ is now completely known:

For $|v| \geq \sqrt{2(U(\xi)-U(x))}$ it is given by φ_+, for $|v| < \sqrt{2(U(\xi)-U(x))}$ it is determined via (40).

Consequently f_+, f_- are determined in $[\xi,x_1] \times \mathbb{R}$ by (36), (39).

We can continue the procedure, e.g., to the right of x_1. $f_+(x_1,v)$, $f_-(x_1,v)$ are known. Therefore $f_+(x,v)$ is determined for $(x,v) \in [x_1,x_2] \times \mathbb{R}$, where x_2 is the point of the next maximum. In that strip f_- can be constructed by solving Abel's integral equation. In that way, we get the solution everywhere.

We have to mention, that we are looking for what is called a "mild" solution of (34), i.e. a solution, which is constant along the characteristics (the level curves) and continuous.
We get the following

<u>Theorem 7</u>: Suppose $E \in C^{(2)}(\mathbb{R})$ such that $E''(x) = O(E(x))$ for $x \to \eta$ for any η with $E(\eta) = 0$, and assume $\varphi_\pm \in C_o \cap L_1$ with the following properties:

(a) If $\Delta_+(\xi) = \sup(U(x)-U(\xi))$, $\Delta_-(\xi) := -\inf(U(x)-U(\xi))$, then $\varphi_\pm$ is even and Hölder continuous for $\frac{v^2}{2} \leq \Delta_\pm(\xi)$

(b) $E'(\xi) = c_- \int\limits_{-\infty}^{+\infty} {}_-dv - c_+ \int\limits_{-\infty}^{+\infty} \varphi_+ dv$

Then the system (34), (35) has a unique mild solution, where E is the given function and $f_\pm(\xi,v) = \varphi_\pm(v)$.

<u>Remark</u>: The assumptions on E and $\varphi_\pm$ are natural and necessary to get a mild solution. This solution can be explicitly constructed and is called a BGK-mode.
There is only one weak point: There are no simple conditions for E (or ρ respectively) and $\varphi_\pm$, such that $f_\pm$ is everywhere non-negative. So one really has to calculate the solutions to be sure that they are physically relevant.

I want to show you another derivation of the solution, which in my opinion is nice and shows something about the increase of complexity in higher dimensions.

In order to simplify the subject we consider again the interval $[x_o,x_1]$, where U is strictly monotonically decreasing, and let $\xi \in [x_o,x_1]$.

Again we assume, that E and f_- are already known in $[\xi,x_1] \times \mathbb{R}$, so we have to solve

$$v \frac{\partial f_+}{\partial x} - E \frac{\partial f_+}{\partial v} = 0$$

with $f_+(\xi,v) = \varphi_+(v)$ and $\int\limits_{-\infty}^{+\infty} f_+(x,v)dv = \rho_+(x)$ given.

Let us forget about characteristics but make a Fourier transform of the equation with respect to v:

With $u(x,y) := \int\limits_{-\infty}^{+\infty} e^{-ivy} f(x,v)dv$, we get formally

$$iu_{xy} - E\,i\,y\,u = 0$$

with $u(\xi,y) = \hat{\varphi}_+(y) = \int\limits_{-\infty}^{+\infty} e^{-ivy}\varphi_+(v)dv$ and $u(x,0) = \rho_+(x)$.

We end up with a characteristic "initial value" problem for a hyperbolic equation:

$$(41) \quad u_{xy}-E(x)yu = 0, \; u(\xi,y) = \hat{\varphi}_+(y), \; u(x,0) = \rho_+(x)$$

There is a simple trick to solve this equation, a trick, which seems not to be well-known: If you have to solve $u_{xy} + f(x)\cdot g(y)u = 0$, substitute

$$u(x,y) = U(F(x),G(y)),$$

where F, G are integrals of f, g respectively. Then you get

$$U_{\xi\eta} + U = 0$$

and the Riemann function for that equation is just

$$G(\xi,\eta;\xi',\eta') = I_o(2\sqrt{(\xi-\xi')(\eta-\eta')}).$$

Therefore, you get as a Riemann function for the original problem

$$(42) \quad G(x,y;x',y') = I_o(2\sqrt{(U(x)-U(x'))(\frac{y^2}{2}-\frac{y'^2}{2})})$$

where I_o is the zeroth order Besselfunction of first kind. G is always real, even if the argument is complex.

The Riemann function G is defined to be a solution of u_{xy}-E(x)yu=0 with respect to the variables x,y and satisfies the boundary conditions

$$G(x',y,x',y') = G(x,y',x',y') = 1 \quad \text{for all } x,y,x',y'.$$

Using (42), one gets the solution of the "characteristic" boundary value problem (41) as

$$u(x,y) = d\, I_0\left(\sqrt{2(U(\xi)-U(x))}\,y\right) + \int_\xi^x I_0\left(\sqrt{2(U(t)-U(x))}\,y\right)\rho_+'(t)\,dt$$

$$+ \int_0^y I_0\left(\sqrt{2(U(\xi)-U(x))(y^2-t^2)}\right)\hat{\varphi}_+'(t)\,dt$$

with $d := u(\xi,0) = \hat{\varphi}_+(0) = \rho_+(\xi)$.

It is possible to calculate explicitly the inverse Fourier transform of u in order to get f_+. For example, the first term yields

$$\frac{d}{\pi}\,\frac{1}{\sqrt{2(U(\xi)-U(x))-v^2}} \quad \text{if } 2(U(\xi)-U(x))-v^2 > 0 \text{ and } 0 \text{ otherwise.}$$

Note the Hamiltonian. This approach yields exactly the same explicit solution of the problem as the method using Abel's integral equation.

What about higher dimensions?
If one is interested in a very special kind of solutions - those depending only on the Hamiltonian, i.e. for example

$$f_-(x,v) = \phi_-(\,\|v\|^2 - 2U(x))$$

everything works like in the one dimensional case; instead of Abel's integral equation, one gets an equation of the form

$$g(z) = \int_y^a k(t)\,(t-y)^{\frac{k-2}{2}}\,dt, \quad \text{where } k=1,2,3$$

is the dimension.
This equation can be easily solved by using fractional derivates of g (for k=2 there is nothing to solve!) and one gets again explicit solutions, the socalled "more-dimensional BGK-modes".
But - in contrast to the one-dimensional case - these solutions are by far not all solutions of the problem - besides the energy, there are more integrals for the Newtonian system, which cannot be given explicitly. For example Horst remarked, that if k=2 and $\tilde{\phi}$ is a stationary solution for the stellardynamic case (for example a two-

dimensional BGK-mode), then

$$\phi(x,v) = \tilde{\phi}(x,v)(1+\Psi(x_1v_2 - x_2v_1))$$

is also a stationary solution, if $\Psi: \mathbb{R} \to [-1,1]$ is an odd function. If one tries to play the Fourier transform trick, one gets for k=2

$$u_{x_1y_1} + u_{x_2y_2} \pm \langle E(x_1,x_2)y\rangle u = 0.$$

This may be transformed to

$$\Delta_\xi \tilde{u} - \Delta_\eta \tilde{u} \pm \langle E(\xi+\eta),\xi-\eta\rangle \tilde{u} = 0, \quad \xi := \frac{x+y}{2}, \quad \eta = \frac{x-y}{2}.$$

The difference of the two Laplacians show that this is a socalled "ultrahyperbolic" equation; not very much is known about that kind of partial differential equations. In particular, it is hard to decide, what kind of boundary value problems are properly posed - remember, that we need $u(x,0) = \tilde{u}(\xi,\xi)$ as data to linearize the problem. The problem of determining all stationary solutions in more than one dimension stays widely open.

Final remarks: Almost all of the material presented in these lectures is mainly of mathematical interest. Now questions of real physical interest arise. f.e.:

1) What is the qualitative character of a global solution? How does the system behave in a long run of time? When is it periodic?

2) Which of the stationary solutions are stable, which are not?

The second question can only be asked with respect to the BGK-modes. A lot of results concerning the problem of linear stability are available, most of them not really rigorous, but they can eventually be made rigorous with some effort. Nothing rigorous is known on non-linear stability as far as I know. But this should be a field of great interest with respect to physical applications, especially since BGK-modes have really been observed in a plasma a few years ago.

Concerning the first question there is almost nothing known today. The only exception is an example given by Kurth, which I want to put on the end of these lectures. We consider the stellardynamic case and are interested in a solution $f(t,\cdot)$, whose spatial density $\rho(t,\cdot)$

is constant in a certain ball whose diameter $r=r(t)$ depends on time

$$\rho(t,x) = \frac{1}{\frac{4\pi}{3}} r^{-3}(t) \qquad \text{if } \|x\| < r(t)$$

and

$$\rho(t,x) = 0 \quad \text{otherwise.}$$

Choose $r(t)$ as a solution of

$$r^3\ddot{r} + r = 1, \; r(0) = 1, \; \dot{r}(0) = H = \text{const.}$$

This solution can be calculated, depending on H. Then define

$$f(t,x,v) = \begin{cases} \frac{3}{2^3}\left[1 - \frac{\|x\|^2}{r^2(t)} - \|r(t)v - \dot{r}(t)x\|^2 + \|x \times v\|^2\right]^{-\frac{1}{2}} & \text{if } \|x\| < r(t) \\ 0 & \text{otherwise} \end{cases}$$

Then f is a solution of the Vlasov-equation, whose spatial density is of the desired form. H might be interpreted as the (dimensionless) Hubble constant. As you may realize there is a lot of astrophysical experience needed to find such a solution. Its behaviour strongly depends on H: For $H < 1$ it behaves periodically, for $H > 1$ the density ρ goes to zero with respect to the $L^\infty(\mathbb{R}^3)$ - Norm.

One might realize by considering this example how complicated the answers for example on questions of the stability or instability of solutions might be.
Nevertheless, working in the field of the Vlasov equation one follows somehow the interest and the efforts of most of the great mathematicians of the 17th and 18th century: Most of the work they did were concerned with astronomy, the explanation of the behaviour of the universe (compare for example the highly interesting book "Mathematics, The Loss of Certainty" by Morris Kline, New York 1980). Therefore, the research on the Vlasov equation could be a challenge even for modern mathematicians.

Acknowledgement. Most of the own research work presented in these lectures was done during the last ten years mainly in collaboration with some of my collegues and students. I want to mention here H. von der Bank, J. Wick, R. Illner and K.H. Petry. The final version of the lectures were worked out during my one month stay at the Centre du recherche pure et appliquée at the Université de Montréal and I want to thank Prof. Anatol Joffe giving me the opportunity to work in the inspiring atmosphere of the centre and especially Prof. Marvin Shinbrot for many fruitful and inspiring discussions.

LITERATURE

For the first lecture:

[1] Braun, W.; Hepp K.: The Vlasov dynamics and its fluctuations in the 1/N limit of interacting classical particles. Comm. Math. Phys. 56 (1977) 101-113.

[2] Dobrushin, R.L.: Vlasov's equation. Functional Anal. and its Appl. 13 (1979) 115-123.

[3] Neunzert, H.: Neuere qualitative und numerische Methoden in der Plasmaphysik. Vorlesungsmanuskript, Paderborn 1975.

[4] Neunzert, H.: Mathematical investigations on particle-in-cell methods. Fluid Dynamics Transactions 9, Warzawa 1978.

[5] Spohn, H.: On the Vlasov Hierarchy. Math. Meth. in the Appl. Sci. 3 (1981).

For the second lecture:

Besides [3] and [4]

[6] Batt, J.: Ein Existenzbeweis für die Vlasov-Gleichung der Stellardynamik bei gemittelter Dichte. Arch. Rational Mech. Anal. 13 (1963) 296-308.

[7] Hannoschöck, G.; Rautmann, R.: Schwache Lösungen und Fehlerabschätzungen für die Vlasov-Gleichung bei gemittelter Dichte. Z. Angew. Math. Mech. 56 (1976) T254-T256.

[8] Horst, E.: Zum statistischen Anfangswertproblem der Stellardynamik. Diplomarbeit München 1975.

[9] Rautmann, R.: Zur globalen Lösbarkeit der Vlasovschen Gleichung für Medien geringer Dichte mit relativistischer Korrektur. Z. Angew. Math. Mech. 48 (1968) T276-T279.

[10] Strampp, W.: On Vlasov's equation with mollified density for an electron gas in an exterior magnetic field. Preprint 1978.

For the third lecture:

[11] Arsen'ev, A.A.: Global existence of a weak solution of Vlasov's system of equations. U.S.S.R. Computational Math. and Math. Phys. 15 (1975) 131-143.

[12] Arsen'ev, A.A.: Existence and uniqueness of the classical solutions of Vlasov's system of equations. U.S.S.R. Computational Math. and Math. Phys. 15 (1975) 252-258.

[13] Batt, J.: Ein Existenzbeweis für die Boltzmann-Vlasov-Gleichung im eindimensionalen Fall. Berichte der Kernforschungsanlage Jülich Jül-126-Ma, 1963.

[14] Batt, J.: Global symmetric solutions of the initial value problem of stellar dynamics. J. Differential Equations 25 (1977) 342-364.

[15] Cooper, J.; Klimas A.: Boundary Value Problem for the Vlasov-Maxwell Equation in One Dimension. Journal of Math. An. and Appl. 75 (1980).

[16] Horst, E.: Zur Existenz globaler klassischer Lösungen des Anfangswertproblems der Stellardynamik. Dissertation München 1979.

[17] Horst, E.: On the existence of global classical solutions of the initial value problem of stellar dynamics. In "Math. Problems in the Kinetic Theory of Gases" (ed. by D.C. Pack and H. Neunzert), 1980.

[18] Horst, E.: On the Classical Solutions of the Initial Value Problem for the Unmodified Non-Linear Vlasov Equation. Math. Meth. in the Appl. Sci. 3 (1981) 229-248.

[19] Iordanskii, S.V.: The Cauchy problem for the kinetic equation of plasma. Amer. Math. Soc. Transl. Ser. 2 35 (1964) 351-363.

[20] Illner, R.; Neunzert, H.: An Existence Theorem for the Unmodified Vlasov Equation. Math. Meth. in the Appl. Sci. 1 (1979) 530-554.

[21] Kurth, R.: Das Anfangswertproblem der Stellardynamik. Z. Astrophys. 30 (1951) 213-229.

[22] Strampp, W.: Special symmetric solutions of Vlasov's equation. Preprint 1978.

[23] Ukai, S.; Okabe, T.: On classical solutions in the large in time of two-dimensional Vlasov's equation. Osaka J. Math. 15 (1978) 245-261.

[24] Wollmann, S.: Global-in-time solutions on the two-dimensional Vlasov-Poisson system. Comm. Pure Appl. Math.

For the fourth lecture:

[25] Neunzert, H.; Petry, K.H.: Ein Existenzsatz für die Vlasov-Gleichung mit selbstkonsistentem Magnetfeld. Math. Meth. in the Appl. Sci. 2 (1980) 429-444.

[26] Petry, K.J.: Zur Existenz von Maßlösungen eines Anfangswertproblems für die gemittelte Vlasov-Gleichung. Dissertation Kaiserslautern 1980.

For the fith lecture:

Besides [3] and [4]

[27] Denavit, J., Walsh, J.M.: "Non-random initialization of particle codes." To appear in: Comm. Plasma Physics and Controlled Fusion.

[28] Kürschner, P.: Numerische Lösung der Vlasov-Gleichung mit finiten Elementen. Dissertation Karlsruhe 1977.

[29] Neunzert, H.; Wick, J.: Die Darstellung von Funktionen mehrerer Veränderlicher durch Punktmengen. Berichte der Kernforschungsanlage Jülich Nr. 996 (1973).

[30] Neunzert, H.; Wick, J.: Die Theorie der asymptotischen Verteilung und die numerische Lösung von Integrodifferentialgleichungen. Num. Math. 21 (1973) 234-243.

[31] Neunzert, H.; Wick, J.: The Convergence of Simulation Methods in Plasma Physics. In "Mathematical Methods of Plasmaphysics" (ed. by R. Kress and J. Wick), 1980.

[32] Wick, J.: Zur Anwendung der Approximation durch endliche Punktmengen auf die Lösung von Integro-Differentialgleichungen. Berichte der Kernforschungsanlage Jülich Nr. 1124 (1974).

For the last lecture:

[33] Bernstein, I.B.; Green, J.M.; Kruskal, M.D.: Exact nonlinear plasma oscillations. Phys. Rev. 108 (1957).

[34] Horst, E.: Private communication.

[35] Kurth, R.: A global particular solution to the initial-value problem of stellar dynamics. Quarterly of Applied Math. Oct. 1978.

[36] Neunzert, H.: Über ein Anfangswertproblem für die stationäre Boltzmann-Vlasov-Gleichung. Berichte der Kernforschungsanlage Jülich Nr. 297 (1965).

THE BOLTZMANN EQUATION AND ITS PROPERTIES

Paul F. Zweifel

"In principio erat verbum."

Ioannes Ii

1. THE BOLTZMANN EQUATION AND ITS PROPERTIES

1. "DERIVATION" OF THE EQUATION.

Throughout these lectures, we shall consider a substance -- generally a gas -- whose density $n(\vec{r},\vec{v},t)$ is a function of position $\vec{r}$, velocity $\vec{v}$, and time t. If a system contains various species of molecules, as, for example air, a subscript "i" will be used to denote the i-th species.

Our first task is to derive, at least heuristically, an equation obeyed by $n(\vec{r},\vec{v},t)$; our second task is to attempt to construct solutions to the equation. Actually, the equation which we shall consider -- the non-linear Boltzmann equation -- is so intractable that, except for a few very special cases, solutions have not been constructed. In fact, even existence of solutions is not known in general. Our primary concern in this series of lectures is to consider the present status of existence theory; we shall sketch the details of such proofs as have been developed, and perhaps -- hopefully -- set the stage for further progress.

Two basic approaches are possible in attempting to derive the Boltzmann equation, i.e., a continuity equation for the single particle density $n(\vec{r},\vec{v},t)$. The first, more mathematical approach, is to integrate the Liouville equation, which describes the N-particle density, over the 6(N-1) variables $(\vec{r}_2 \ldots \vec{r}_N, \vec{v}_2 \ldots \vec{v}_N)$. Unfortunately, this does not lead to a single equation for $n(\vec{r},\vec{v},t)$, but rather to a coupled hierarchy of equations (known as the BBGKY-hierarchy) in which the equation for the single-particle density $n(\vec{r},\vec{v},t)$ involves the two particle density, the equation for the two particle density involves the three particle density, and so on, ad infinitum. One can obtain kinetic equations, i.e., equations for $n(\vec{r},\vec{v},t)$ alone, by making various assumptions on the form of the two particle density. One then has the problem of attempting to assess the validity and/or range of applicability of these assumptions, which is a formidable task in itself.

Since the main purpose of these lectures is to discuss the existence of solutions to the Boltzmann equation, rather than its derivation, we shall adopt a more heuristic approach, based on physical considerations. (This approach has at least historical merit, since it was the method employed by Boltzmann in his original "derivation".)[1]

Consider the balance of gas molecules in a small volume element $\Gamma x\Delta$ at $\vec{r},\vec{v}$ in phase space. (Γ is the element of velocity space and Δ of configuration space.) Recalling that n is a density, we can write

$$\frac{\partial}{\partial t}\int_{\Gamma x\Delta} d^3r d^3v n(\vec{r},\vec{v}) = \text{"Production"} - \text{"Destruction" in } \Delta. \tag{1}$$

The "Production" term involves gas molecules which suffer a collision at some velocity $\vec{v}'$ and emerge with the velocity $\vec{v}$. The "Destruction" term includes those molecules which stream freely out of Δ without collision, plus those molecules of velocity $\vec{v}$ which suffer a collision in Δ and emerge with a new velocity $\vec{v}'$. Let us write expressions for these three components of the right hand side of Eq. (1):

$$\text{Loss due to streaming} = \int_{\Gamma} d^3v \int_{\partial\Delta} n\vec{v}\cdot d\vec{s} \tag{2}$$

since $n\vec{v}$ is the current density.

The loss due to scattering is proportional to the collision rate between pairs of molecules. A typical collision is shown in Fig. I, in which two molecules of speed $\vec{v}$ and $\vec{v}_1$ collide, emerging with velocities $\vec{v}'$, $\vec{v}_1'$. Conservation of momentum and energy require that in every collision

$$\vec{v} + \vec{v}_1 = \vec{v}' + \vec{v}_1' \tag{3a}$$

and

$$v^2 + v_1^2 = v'^2 + v_1'^2 \tag{3b}$$

assuming all molecules have the same mass (otherwise, Eqs. (3) would include individual masses).

The collision rate, between molecules of velocity $\vec{v}$ and $\vec{v}_1$ is inversely proportional to the mean time between collisions. If λ represents the collision mean free path, this mean time is given by $\lambda/|\vec{v}-\vec{v}_1|$ since $|\vec{v}-\vec{v}_1|$ is the relative distance the molecules travel each second.

$$\text{Collision rate} \propto \frac{|\vec{v}-\vec{v}_1|}{\lambda} . \tag{4}$$

It is customary to write $1/\lambda$ as the integral of a differential cross section $\sigma(\vec{v},\vec{v}_1 \to \vec{v}',\vec{v}_1')$ times the density $n(\vec{r},\vec{v}_1,t)$. This, in fact, defines the cross section.

We now note that the conservation laws, Eqs. (3), along with the (presumed) rotational invariance of the scattering process imply that the cross-section σ depends only $v_{rel} = |\vec{v}-\vec{v}_1| = |\vec{v}'-\vec{v}_1'|$ and $\mu = \vec{v}\cdot\vec{v}'/vv' = \vec{v}_1\cdot\vec{v}_1'/v_1v_1'$. Then

$$2\pi\sigma(v_{rel},\mu) = \sigma(\vec{v},\vec{v}_1 \to \vec{v}',\vec{v}_1')\left(\frac{d^3v'}{d\mu}\right) \tag{5a}$$

where the Jacobian obviously has the property

$$\left(\frac{d^3v'}{d\mu}\right) = \left(\frac{d^3v}{d\mu}\right) . \tag{5b}$$

Then

$$\text{Collision rate} \propto \int d^3v_1 d^3v' v_{rel}\ \sigma(\vec{v},\vec{v}_1 \to \vec{v}',\vec{v}_1')n(\vec{r},\vec{v}_1,t) \tag{6}$$

where the integrals insure that all possible contributions are summed. Finally, the total collision rate is given by the expression in Eq. (6) multiplied by the number of molecules of velocity $\vec{v}$ within $\Gamma\times\Delta$:

"Destruction" due to collisions =

$$= \int_{\Gamma\times\Delta} d^3v d^3rn(\vec{r},\vec{v},t) \int d^3v_1 d^3v' v_{rel}\sigma(\vec{v},\vec{v}_1 \to \vec{v}',\vec{v}_1')n(\vec{r},\vec{v}_1,t). \tag{7}$$

Combining Eqs. (2) and (7), and using the divergence theorem yields the total destruction term:

$$\text{"Destruction"} = \int_{\Gamma\ \Delta} d^3r d^3v\{\vec{\nabla}\cdot\vec{v}n(\vec{r},\vec{v},t) + n(\vec{r},\vec{v},t) \int d^3v_1 d^3v' v_{rel}\sigma(\vec{v},\vec{v}_1 \to \vec{v}',\vec{v}_1')n(\vec{r},\vec{v}_1,t)\}. \tag{8}$$

The "Creation" term can be obtained by arguments completely analogous to those used in arriving at Eq. (7). One finds

$$\text{"Creation"} = \int_{\Gamma\times\Delta} d^3r d^3v \int d^3v_1' d^3v' v_{rel} \sigma(\vec{v}', \vec{v}_1' \to \vec{v}, \vec{v}_1) n(\vec{r}, \vec{v}', t) n(\vec{r}, \vec{v}_1', t). \tag{9}$$

Recall now that

$$\left(\frac{d^3v}{d\mu}\right) = \left(\frac{d^3v'}{d\mu}\right).$$

Furthermore, Liouville's Theorem implies that $d^3vd^3v_1 = d^3v'd^3v_1'$. Using these facts along with Eq. (5a) we can combine the creation and destruction terms into a single integral so that Eq. (1) can be written in the compact form

$$\int_{\Gamma\times\Delta} d^3vd^3r\{\frac{\partial n(\vec{r},\vec{v},t)}{\partial t} + \vec{v}\cdot\vec{\nabla}n - J(n,n)\} = 0 \tag{10}$$

where the "collision integral" J(n,n) is given by

$$J(n,n) = 2\pi\int d^3v_1 \int d\mu v_{rel} \sigma(v_{rel},\mu)[n(\vec{r},\vec{v}',t)n(\vec{r},\vec{v}_1',t) - \\ - n(\vec{r},\vec{v},t)n(\vec{r},\vec{v}_1 t)]. \tag{11}$$

Since the phase space element $\Gamma\times\Delta$ is arbitrary, the integrad of Eq. (10) must vanish identically. The non-linear Boltzmann equation then becomes

$$\frac{\partial n(\vec{r},\vec{v},t)}{\partial t} + \vec{v}\cdot\vec{\nabla}n(\vec{r},\vec{v},t) = J(n,n), \tag{12}$$

with J(n,n) given by Eq. (11).

It is important to stress that the analysis leading to Eq. (12) cannot be considered a derivation in any strict sense. Rather, it is a plausibility argument, and the Boltzmann equation may not describe any kind of physical reality. In fact, as we shall see in the course of these lectures, except in special cases, the question of existence of solutions to the Boltzmann equation is still open. In particular, several assumptions have been introduced implicitly in the preceeding analysis. They are

(0) The molecules have no internal degrees of freedom, all collisions are elastic.

(1) The gas is sufficiently dilute that only binary collisions need be considered.

(2) There are no external force fields present (actually, this assumption can be relaxed rather easily).

(3) The velocity $\vec{v}$ of gas molecule is uncorrelated with the position $\vec{r}$ (Boltzmann's Stosszahl Ansatz).

In particular, assumption (3) allowes us to write the collision rate as proportional to the independent densities $n(\vec{v})n(\vec{v}_1)$.

2. EQUILIBRIUM DISTRIBUTION AND COLLISION INVARIANTS

The Boltzmann equation has the main purpose of describing systems which are not in thermodynamic equilibrium. However, a description of the equilibrium state is actually hidden in the equation, in the following sense. In true equilibrium. n is independent of time t, in which case Eq. (12) reduces to

$$\vec{v}\cdot\vec{\nabla}n_0(\vec{r},\vec{v}) = J(n_0,n_0) = \int d^3v_1 \int d\mu v_{rel}\sigma(v_{rel},\mu)(n_0(\vec{r},\vec{v}')n_0(\vec{r},\vec{v}_1') - n_0(\vec{r},\vec{v})n_0(\vec{r},\vec{v}_1)) \tag{13}$$

where we have used the symbol n_0 to denote the equilibrium distribution. Then a sufficient condition that a solution exists is that n_0 be independent of $\vec{r}$ and that

$$n_0(\vec{v}')n_0(\vec{v}_1') = n_0(\vec{v})n_0(\vec{v}_1). \tag{14a}$$

$$\text{or } \log n_0(\vec{v}) + \log n_0(\vec{v}_1) = \text{constant} \tag{14b}$$

Eq. (14b) may be considered a conservation law.[2] In particular let $\chi_i(\vec{v})$ represent a conserved quantity. That is, in a collision $\chi_i(\vec{v}) + \chi_i(\vec{v}_1) = \chi_i(\vec{v}') + \chi_i(\vec{v}_1')$. Then

$$\log n_0(\vec{v}) = \sum_i \alpha_i \chi_i(\vec{v}) \tag{15}$$

where the sum is taken over all conserved quantities, and the α_i are (real) constants. But, neglecting spin, there are only five such invariant quantities, namely

$$\chi_0 = C \tag{16a}$$

$$\vec{\chi}_1 = \vec{v} \tag{16b}$$

$$\chi_2 = v^2 \tag{16c}$$

(Here C is a constant). Except for multiplicative factors, $\vec{\chi}_1$ represents the momentum and χ_2 the energy. Thus

$$\log n_0(\vec{v}) = \alpha_1 C + \vec{\alpha}_2 \cdot \vec{v} + \alpha_3 v^2 \tag{17a}$$

or, for convenience, defining new constants K, A and $\vec{v}_0$

$$\log n_0(\vec{v}) = \log K - A(\vec{v}-\vec{v}_0)^2 \tag{17b}$$

so that

$$n_0(\vec{v}) = Ke^{-A(\vec{v}-\vec{v}_0)^2}. \tag{18}$$

To determine the constants K, A and $\vec{v}_0$ we have the condition

$$\int d^3v n_0(\vec{v}) = \rho, \tag{19}$$

where ρ is the particle density. Also if $\vec{v}$ is the average velocity, we see

$$\vec{v} = \int d^3v \vec{v} n_0(\vec{v}) / \int d^3 v n_0(\vec{v}) \tag{20}$$

while the average energy ε is

$$\varepsilon = \frac{3}{2}k_B T = \int d^3v \tfrac{1}{2}mv^2 n_0(\vec{v}) / \int d^3v n_0(\vec{v}) \tag{21}$$

where k_B is Boltzmann's constant and T the temperature.

The five equations (19-(21) may be used to determine the constants K, A and $\vec{v}_0$. One finds

$$n_0(\vec{v}) = \rho \left(\frac{m}{2\pi kT}\right)^{3/2} e^{-\frac{m}{2k_B T}(\vec{v}-\vec{v}_0)^2} \tag{22}$$

Since Eq. (14a) was a sufficient condition for equilibrium, Eq. (22) represents a <u>possible</u> equilibrium distribution. We shall show in the next section that it is, indeed, the only possible equilibrium. Eq. (14a) is also a <u>necessary</u> condition.

The average velocity $\vec{v} = \vec{v}_0$ is sometimes called the "drift velocity". If the gas has no mass motion (the typical case) $\vec{v}_0 = 0$. Eq. (22) is, of course, the celebrated Maxwellian distribution, and is seen to be the equilibrium distribution described by the Boltzmann equation. The quantities χ_0, $\vec{\chi}_1$ and χ_2 are often called "collision invariants".

If an external force field $\vec{F}$ is present (cf. assumption (2) supra) Eq. (22) is modified to read

$$n_0(\vec{r},\vec{v}) = \rho \left(\frac{m}{2\pi k_B T}\right)^{3/2} e^{-\beta(\frac{m}{2}(\vec{v}-\vec{v}_0)^2 + \Phi(\vec{r}))} \tag{23}$$

where $\vec{F} = -\vec{\nabla}\Phi$ and we have introduced the standard notation $\beta = 1/k_B T$. Actually, we shall not consider such force fields in these lectures.

3. BOUNDARY CONDITIONS; THE H-THEOREM AND COLLISION INVARIANTS.

The full spatially-dependent Boltzmann equation, Eq. (12), requires boundary conditions be imposed on $\partial\Delta$ where the gas molecules undergo reflections. We shall consider two possible types of reflection, specular reflection and diffuse reflection. In specular reflection (Latin: speculum-a "mirror") the gas molecules rebound elastically from the wall according to the usually law of reflection "angle of incidence = angle of reflection." In diffuse reflection, the molecule is assumed to be absorbed by the wall, and to be re-emitted in a Maxwell-Boltzmann distribution characterized by the wall temperature, T_w. To express the law of reflection mathematically, consider a point on the surface $\vec{r}_s$, and resolve the molecule's velocity $\vec{v}$ into components $(\vec{v}_{\|}, v_{\perp})$ where v is perpendicular to the wall at $\vec{r}_s$. Then the boundary condition is written

$$n(\vec{r}_s,\vec{v}_{\|}, v_{\perp},t) + (1-\alpha)n(\vec{r}_s,\vec{v}_{\|},-v_{\perp},t) + \alpha M(v,T_w,t), \quad v_{\perp} < 0, \tag{24a}$$

where the Maxwellian $M(v,T_w,t)$ has the same flux as the incoming distribution:

$$\int_{v_{\perp}<0} v_{\perp} M(v,T_w,t)d^3v = \int_{v_{\perp}>0} v_{\perp}\, n(\vec{r}_s,\vec{v},t)d^3v. \tag{24b}$$

That is, a fraction $(1-\alpha)$ of the molecule undergoes specular reflection and a fraction α undergoes diffuse reflection; α is called the "accommodation coefficient."

The boundary condition (24) will be used in order to "prove" the famous H-theorem of Boltzmann. (We have written "prove" in quotation marks because a rigorous proof, from the mathematical point of view, requires a careful study of the existence of the various integrals which we shall introduce. We defer such technical questions to a later lecture, and present now a heuristic (i.e., physicist's) proof.

The quantity H is defined as:

$$H(t) = \int d^3r d^3v \, n(\vec{r},\vec{v},t) \ln n(\vec{r},\vec{v},t) \tag{25}$$

where the d^3r integration is carried out over the total volume of the system under consideration. We shall show that H is a non-increasing function of t. This suggests that H is connected into the thermodynamic entropy S, which, we recall, is a non-decreasing function. In fact

Proposition. $H = C - \frac{1}{k_B} S$, where C is a constant, k_B is Boltzmann's constant and S is the (combinatorial entropy.

Proof.

Suppose phase space is divided into cells $\{\Omega_k\}_{k=1}^{M}$. Then, the number of ways, g, that N molecules can be distributed among the n cells with n_k molecules in the k-th cell is

$$g = \frac{N!}{\prod_{k=1}^{M} n_k!} \tag{26}$$

The entropy S is given by

$$S = k_B \ln g \tag{27a}$$

$$= k_B \ln N! - k_B \sum_k \ln n_k! \tag{27b}$$

Expressing $\ln n_k! = n_k \ln n_k$ (Stirling's approximation) we obtain

$$S = C - k_B \sum_k \ln n_k \tag{28a}$$

$$= C - k_B \int n(\vec{r},\vec{v},t) \ln n(\vec{r},\vec{v},t) d^3r d^3v, \tag{28b}$$

proving the proposition.

We now are in a position to state and give a heuristic proof, of the H-theorem. First, we observe that

$$\frac{dH}{dt} = \int d^3r \int d^3v \frac{\partial n(r,v,t)}{\partial t} [1 + \log n(\vec{r},\vec{v},t)]. \tag{29}$$

Thus, $\frac{\partial n}{\partial t} = 0$, i.e., equilibrium, $\Rightarrow \frac{dH}{dt} = 0$. We now state

<u>H-Theorem (Boltzmann)</u>. Let n satisfy the Boltzmann equation, (12) and H be given by Eq. (25). Then $\frac{dH}{dt} \leq 0$.

<u>Proof</u>. Use Eq. (12) to rewrite (29) as

$$\frac{dH}{dt} = \int d^3r \int d^3v [-\vec{v}.\vec{\nabla}n + J(n,n)][1 + \log n]. \tag{30}$$

Noting that

$$\vec{\nabla}(n \ln n) = \vec{\nabla}n(1 + \ln n) \tag{31}$$

and using the divergence theorem, Eq. (30) can be written

$$\frac{dH}{dt} + \oint\int d^3v n(\vec{r}_s,\vec{v},t) \ln n(\vec{r}_s,\vec{v},t)\vec{v}\cdot d\vec{S} = \int d^3r d^3v J(n,n)[1 + \ln n]. \tag{32}$$

Here the integral $\oint$ is carried out over the total surface enclosing the gas: $d\vec{S}$ is, of course, the outward normal at point $\vec{r}_s$, so that $\vec{v}\cdot d\vec{S} = v_\perp d\vec{S}$, where $v_\perp$ was defined previously.

For simplicity, assume first that the gas is homogeneous, and thus infinite in extent, so that the surface term is not present. Then Eq. (30) reduces to

$$\frac{dH}{dt} = \int d^3r d^3v J(n,n)][1 + \log n], \tag{33a}$$

or

$$\frac{dH}{dt} = 2\pi\int d^3r d^3v d^3v_1 d\mu v_{rel}\sigma(v_{rel},\mu)[n(\vec v')n(\vec v_1')-n(\vec v)n(\vec v_1)][1 + \log n] \tag{33b}$$

where we have written $n(\vec v)$ for $n(\vec r,\vec v,t)$. The integral in Eq. (33) is invariant under the interchange $\vec v\leftrightarrow\vec v_1$ recalling the $v_{rel} = |\vec v-\vec v_1|$ and $\mu = \vec v\cdot\vec v'/vv' = \vec v_1\cdot\vec v_1'/v_1v_1'$. Thus

$$\frac{dH}{dt} = \pi\int d^3r d^3v d\mu v_{rel}\sigma(v_{rel},\mu)(n(\vec v')n(\vec v_1')-n(\vec v)n(\vec v_1))(2 + \log n(\vec v)n(\vec v_1))$$

We now make the change $(\vec v,\vec v_1)\leftrightarrow(\vec v',v_1')$, which also leaves the integral invariant since, as we recall, $|\vec v-\vec v_1| = |\vec v'-\vec v_1'|$,

$\sigma(\mu) = \sigma(\vec v,\vec v_1 \to \vec v',\vec v_1')\left(\frac{d^3v'}{d\mu}\right) = \sigma(\vec v',\vec v_1' \to \vec v,\vec v_1)\left(\frac{d^3v}{d\mu}\right)$ and $d^3v d^3v_1 = d^3v' d^3v_1'$. Thus, also using the trick of interchanging $\vec v'\leftrightarrow\vec v_1'$, we can write

$$\frac{dH}{dt} = -\frac{\pi}{2}\int d^3v d^3v_1 d\mu v_{rel}\sigma(v_{rel},\mu)(n(\vec v')n(\vec v_1')-n(\vec v)n(\vec v_1)) \times$$
$$(\log n(\vec v')n(\vec v_1') - \log n(\vec v)n(\vec v_1)). \tag{34}$$

These formal manipulations have left us with an expression which is manifestly negative, for $v_{rel}\sigma$ is positive and we observe that for $x,y>0$ $(x-y)(\log x-\log y) = y(\frac{x}{y} - 1)\log\frac{x}{y} > 0$. Further, equality holds for only $x=y$, *i.e.* $n(\vec v')n(\vec v_1')-n(\vec v)n(\vec v_1) = 0$. This completes the "proof" for the spatially homogeneous case.

We can generalize this to the spatially dependent case if we can show that the second term on the left hand side of Eq. (32) is non-negative. This is trivial for the case of pure specular reflection ($\alpha=0$ in Eq. (24)), because inserting:

$n(\vec r_s,\vec v_{\|},v_\perp,t) = n(\vec r_s,\vec v\ ,-v_\perp,t)$, $v_\perp< 0$, in the surface integral, that integral is seen to vanish identically (the integrand is an odd function of $v_\perp$). For $\alpha \neq 0$, the situation is somewhat more complicated. In fact, the theorem is not true unless one also considers the entropy change of the wall due to the diffuse reflection.[3] That is, Eq. (32) must be modified to

$$\frac{dH}{dt} + \oint\int d^3v n(\vec{r}_s,\vec{v},t)\ln n(\vec{r}_s,\vec{v},t)v_\perp dS - \frac{1}{k_B}\left(\frac{dS}{dt}\right)_{wall} = -G \tag{35}$$

where -G is the manifestly negative quantity given on the right hand side of Eq. (34). Calling the surface integral in Eq. (35) J and using Eq. (24) and the convexity of xlnx,

$$J \geq \alpha\oint dS\int_{v_\perp>0} d^3v_\perp n(\vec{r}_s,\vec{v},t)\ln n(\vec{r}_s,\vec{v},t)$$
$$+ \alpha\oint dS\int_{v_\perp>0} d^3vv_\perp\{n(\vec{r}_s,\vec{v},t)\ln n(\vec{r}_s,\vec{v},t) - M(v,T_w,t)\ln M(v,T_w,t)\} \tag{36}$$

Furthermore, from a well-known thermodynamic identity,

$$-\left(\frac{dS}{dt}\right)_{wall} = -\frac{1}{T_w}\Delta Q = \frac{\alpha}{T_w}\int_{v_\perp>0}\tfrac{1}{2}mv^2[n(\vec{r}_s,\vec{v},t) - M(v,T_w,t)]v_\perp d^3v, \tag{37}$$

since W, the work done by the wall, is zero.

Combining Eqs. (36) and (37) and writing, for convenience

$$n(\vec{r}_s,\vec{v},t) = hM$$

where we have suppressed the arguments of M

$$J - \frac{1}{k_B}\left(\frac{dS}{dt}\right)_{wall} = \alpha\oint dS\int_{v_\perp>0} v_\perp M[h\ln hM - \ln M + \frac{mv^2}{2k_BT}(h-1)]. \tag{38}$$

Let us write

$$M = Ke^{-\frac{mv^2}{2k_BT}} \tag{39a}$$

so that

$$\ln M = \ln K - \frac{mv^2}{2k_BT} \tag{39b}$$

where K is a normalization constant. Then

$$J - \frac{1}{k_B}\left(\frac{dS}{dt}\right)_{wall} = \alpha\oint dS\int_{v_\perp>0} v_\perp M[h\ln h - (1-h)\ln K]d^3v \tag{40a}$$

$$= \alpha\oint dS\int_{v_\perp>0} v_\perp M[h\ln h + (1-h)]d^3v \tag{40b}$$

since [Eq. (24b)]

$$\int d^3vv_\perp M(1-h) \equiv 0. \tag{41}$$

Now we have the following result, called the Stückelberger-Pauli identity:[4]

$\forall h>0$,

$$P(h) = h\ln h + (1-h) \geqslant 0;\; P(h) = 0 \text{ iff } h=1. \tag{42}$$

[Proof: $P(0) = 0$, $P(1) = 0$, and $P'(0) = \ln h$.]

Thus, the right hand side of Eq. (41) is non-negative, and, from Eq. (35) we calculate $\frac{dH}{dt} \leqslant 0$, proving the "H-theorem."

Remarks: 1. Crucial to the "proof" was the condition $\sigma(\vec{v},\vec{v}_1 \to \vec{v}',\vec{v}_1')\left(\frac{d^3v'}{d\mu}\right) = \sigma(\vec{v}',\vec{v}_1' \to \vec{v},\vec{v}_1)\left(\frac{d^3v}{d\mu}\right)$ or so-called microscopic irreversibility. Without this property of nature, systems would not drive toward equilibrium.

2. We have written "proof" in quotation marks because certain mathematical details have been completely ignored, for example, the existence of integrals defining H. To be precise, one should define a function space in which the solution exists, and show the existence of H. We return to this question later.

COROLLARY. The Maxwell-Boltzmann distribution (22) is the unique solution of Eq. (13), i.e., is the unique equilibrium distribution function. Further, the gas temperature $T = T_w$ the wall temperature.

PROOF. We have already shown that the Maxwell-Boltzmann distribution function is a solution of (13) and that $\frac{\partial n}{\partial t} = 0 \Rightarrow \frac{dH}{dt} = 0$. But from the proof of the H-Theorem, we observe that $\frac{dH}{dt} = 0 \Rightarrow n(\vec{v})n(\vec{v}_1) - n(\vec{v}')n(\vec{v}_1') = 0 \Rightarrow n = n_0$ and that $h = 1$ [Eq. (44)] since surface and volume integrals must vanish independently. Thus $n(\vec{v}) = M(v,T_w)$.

Thus, all systems approach a Maxwell-Boltzmann distribution, independent of the cross section σ, so long as $\sigma>0$ (and of course, the condition of microscopic irreversibility is obeyed). If $\sigma=0$, clearly no collision occurs to "thermalize" a system.

We have already defined the collision invariants $\chi_i(\vec{v})$ (cf. Eq. (16)). We state:

PROPOSITION. Let χ_i be a collision invariant. Then $\int \chi_i(\vec{v}_1)J(n,n)d^3v = 0$.

PROOF. The proof is similar to that of the H-theorem, in that in the integral in Eq. (33) is rewritten three times under changes of variable $\vec{v} \leftrightarrow \vec{v}_1$, $\vec{v},\vec{v}_1 \leftrightarrow \vec{v}',\vec{v}_1'$, and $\vec{v},\vec{v}_1 \leftrightarrow \vec{v}_1',\vec{v}'$. In each case, a different formal expression for the same integral is obtained. Adding them all together and dividing by four yields

$$\int \chi_i(\vec{v}) J(n,n) d^3v = \tfrac{1}{4}\int d^3v d^3v_1 d\mu v_{rel}\sigma(v_{rel},\mu)(n(\vec{v}')n(\vec{v}_1')-n(\vec{v})n(\vec{v}_1)) \times$$
$$(\chi_i(\vec{v}) + \chi_i(\vec{v}_1) - \chi_i(\vec{v}') - \chi_i(\vec{v}_1')) \tag{43}$$

But the integral of Eq. (43) vanishes identically since by definition of a collision invariant

$$\chi_i(\vec{v}) + \chi_i(\vec{v}_1) = \chi_i(\vec{v}') + \chi_i(\vec{v}_1').$$

This proposition implies the existence of conservation laws for the Boltzmann equation. For example, the total number of particles, N is given by

$$N = \int d^3r d^3v n(\vec{r},\vec{v},t) \tag{44}$$

and we expect that N should be constant. Integrating Eq. (12) we obtain

$$\frac{\partial N}{\partial t} + \int d^3v d^3r \vec{v}\cdot\vec{\nabla} n(\vec{r},\vec{v},t) = 0 \tag{45}$$

where the right hand side vanishes by virtue of the proposition. Arguing, as in the proof of the H-theorem, that the integral of the gradient term vanishes, we conclude that $\frac{\partial N}{\partial t} = 0$. Similarly, it easily follows that $\frac{\partial p_i}{\partial t} = 0$ where the momentum $\vec{p}$ is given by

$$p_i = \int v_i n(\vec{r},\vec{v},t) d^3r d^3v \tag{46}$$

and $\frac{\partial E}{\partial t} = 0$ with the energy E defined by

$$E = \int \tfrac{1}{2}mv^2 n(\vec{r},\vec{v},t) d^3r d^3v. \tag{47}$$

Actually, the conservation laws have more content. For consider Eq. (3), with the integration carried out not over all $\vec{r}$ but only the small volume element Δ. We obtain

$$\frac{\partial\rho(\vec{r},t)}{\partial t} + \vec{\nabla}\cdot\vec{j}(\vec{r},t) = 0 \tag{48}$$

with

$$\rho(\vec{r},t) = \int d^3v n(\vec{r},\vec{v},t) \tag{49a}$$

and

$$\vec{j}(\vec{r},t) = \int d^3v \vec{v} n(\vec{r},\vec{v},t). \tag{49b}$$

Similar continuity equations are obtained after multiplying the Boltzmann equation by $\vec{\chi}_1$ and χ_2 and integrating. In this way, the hydrodynamic approximation to the Boltzmann equation is obtained. (The hydrodynamic equations are approximate because the set of moment equations obtained in this way must be truncated if it is to be solved.) We shall not consider the hydrodynamic equations further in these lectures.

"We will now discuss in a little more detail the struggle for existence."
Charles Darwin, The Descent of Man

II. THE SPACE INDEPENDENT EQUATION

4. COLLISION MODELS.

Throughout this chapter we shall assume that the density $n(\vec{r},\vec{v},t)$ is independent of $\vec{r}$, in which case the Boltzmann Eq. (12) reduces to

$$\frac{\partial n(\vec{v},t)}{\partial t} = J(n,n). \tag{50}$$

We shall seek solutions in some Banach space X, the precise nature of the space depending upon the problem being considered; typical space might be $L^1(R^3)$ or $L^\infty(R^3)$. In every case, J: $X \times X \to X$ is a bounded bilinear form. However, since J is the difference of two positive forms (c.f. Eq. (11)), we shall introduce the further requirement that each component of J be finite. This requires the introduction of a cutoff, first used by H. Grad.[5]

It is convenient at this point to adopt some of Grad's notation. To this end, we write the collision integral in the form

$$J(n,n) = \int B(\theta,V)(n'n_1' - nn_1)d\theta d\varepsilon d^3v_1 \tag{51}$$

where now $\vec{V}$ is written for $\vec{v}_{rel}$ $(\vec{V} = \vec{v}_1 - \vec{v})$, $n' \equiv n(\vec{r},\vec{v}',t)$, $n_1' \equiv n(\vec{r},\vec{v}_1',t)$, etc. The angles θ and ε are respectively the polar and azimuthal angles of $\vec{v}'$ in a spherical coordinate system attached to the molecule $(\vec{v})$[6] with z-axis oriented in the direction $\vec{V}$. Then θ is related to the center-of-mass scattering angle $\cos^{-1}\mu \equiv \phi$ introduced in Chapter I by $\phi = \pi - 2\theta$ or to the Laboratory System scattering angle θ_L by $\theta = \frac{\pi}{2} - \theta_L$. Thus $0 \leqslant \theta \leqslant \frac{\pi}{2}$ (recall[7] that in an elastic scattering collision between two identical particles, the maximum deflection in the Laboratory Coordinate System is 90°).

The quantity $B(\theta,V)$ is, of course, related to the differential scattering cross section. From the dynamics of collision processes,[8] B can be expressed in terms of the so-called impact parameter r, defined to be the perpendicular distance from $(\vec{v})$ to the orbit of $(\vec{v}_1)$ when the latter is far away:

$$B(\theta,V) = Vr\left|\frac{\partial r}{\partial \theta}\right|. \tag{52}$$

Thus, the total cross section σ is given by

$$V\sigma = \int B(\theta,V)d\theta d\varepsilon \tag{53a}$$

$$= V \int r dr d\varepsilon. \tag{53b}$$

Eq. (23) makes it clear why a cutoff is needed. If the impact parameter can take on any value from zero to infinity, the total cross section is infinite. However, under the reasonable assumption that the range of the inter-molecular force is finite, with range $r_0 < \infty$, say, then we have $r \leqslant r_0$, and the cross section is finite. This "cutoff" can, by virtue of Eq. (52), also be thought of as an angular cutoff on $B(\theta, V)$. Specifically, σ will be finite if

$$\exists \theta_0 > 0 \text{ s.t. } B(\theta,V) = 0 \text{ for } \theta < \theta_0.$$

Thus, the assumption of a finite range force in the Boltzmann gas corresponds to the absence of grazing collisions.

We hereafter assume, without further comment, that we are dealing with a "cutoff" $B(\theta, V)$, whatever scattering model we shall consider. In fact, we shall be primarily interested in two models:

A) HARD SPHERE MODEL. We assume, in this model, that collisions consist of elastic billiard ball type events between molecules of radius r_0. Then[5]

$$B(\theta,V) = r_0^2 V \sin\theta \cos\theta \tag{54a}$$

so that

$$\sigma = \pi r_0^2. \tag{54b}$$

B. INVERSE POWER REPULSION. If the intermolecular force is inversely proportional to some power of the separation distance:

$$F = K/|\vec{r} - \vec{r}_1|^s, \quad s > 3 \tag{55}$$

one can calculate

$$B(\theta, V) \propto V^{\frac{s-5}{s-1}} \Lambda(\theta); \tag{56}$$

that is, the cross section factors into a power of V times a function of θ. We shall be mainly interested in the case of the so-called "Maxwell molecule" corresponding to s=5. In this case, B is independent of V, and we shall simply write $B(\theta)$ for this case. (To be precise, we shall consider "pseudo-Maxwell molecules," i.e. a Maxwell molecule with the Grad angular cutoff incorporated into $B(\theta)$.)

5. EXISTENCE THEORY FOR THE PSEUDO-MAXWELL GAS (SPATIALLY HOMOGENEOUS).

We wish to consider the existence of solutions to the initial value problem

$$\frac{\partial n}{\partial t} = J(n,n) \tag{57a}$$

$$n(\vec{v},0) = f_0(\vec{v}). \tag{57b}$$

We seek solutions in the Banach space $L^1(R^3)$. It is convenient to decompose J as

$$J(n,n) = L(n,n) - \nu(n)n \tag{58}$$

with

$$\nu(n) = \int B(\theta)n_1 d\theta d\varepsilon d^3v_1 = V\sigma\rho \tag{59a}$$

and

$$L(n,n) = \int B(\theta)n'n_1' d\theta d\varepsilon d^3v_1. \tag{59b}$$

L is a bounded form, L: $L^1(R^3) \times L^1(R^3) \to L^1(R^3)$ since

$$\|L(n,n)\|_{L^1} = \int d^3v \int B(\theta)n'n_1' d\theta d\varepsilon d^3v_1 \tag{60a}$$

$$= \int B(\theta)d\theta d\varepsilon \int d^3v' d^3v_1' n'n_1' \tag{60b}$$

$$= V\sigma\rho^2. \tag{60c}$$

Here, we have used the fact (cf. Chapter I) that $d^3vd^3v = d^3v'd^3v_1'$, and have introduced the density, ρ, defined in Eq. (19). Note $V\sigma$ is independent of V.

The standard procedure for proving existence theorems for non-linear equations[9] is to convert the equation into an integral equation, set up an iteration scheme for the solution and prove convergence. The existence of a limit to the iteration scheme guarantees that a solution to the integral equation exists. Such a solution is sometimes called a "mild solution" to the original equation. If it can be shown that the mild solution is differentiable, it follows that the mild solution is also a "strong" solution, i.e. a solution to the original equation. Generally, given a mild solution, further requirements are necessary on the initial data in order to guarantee the existence of a strong solution, e.g. continuity (in the Banach-space norm).

The density ρ, we recall, is a constant, independent of t, since integration of Eq. (57a) over d^3v gives

$$\frac{\partial \rho}{\partial t} = \int J(n,n)d^3v = 0, \tag{61}$$

where Eq. (30) has been utilized. Thus, Eq. (57a), which can be rewritten as

$$\frac{\partial n}{\partial t} + \nu n = L(n,n) \tag{62}$$

can be converted into the integral equation

$$n(\vec{v},t) = e^{-\nu t} f_0(\vec{v}) + \int_0^t e^{-\nu(t-\tau)} L(n,n)(\tau)d\tau. \tag{63}$$

Define an iteration scheme by

$$n_{k+1} = f_0 e^{-\nu t} + \int_0^t L_k e^{-\nu(t-\tau)} d\tau \tag{64a}$$

$$L_k = \int B(\theta) n_k' n_{1,k}' d\theta d\varepsilon d^3v_1, \tag{64b}$$

with $n_0 = 0$. We now state:

<u>THEOREM I</u>. The iteration scheme (64) converges to a function $n \in L^1(R^3)$ which is thereby a mild solution to Eq. (62).

Before proving the theorem, we point out that $L(n,n)$ is continuous, i.e. for a sequence $\{f_k\}$ with $f_k \to f$, $L(f_k,f_k) \to L(f,f)$. where the limits are taken in the L^1 norm. While the equivalence between boundedness and continuity is well-known for linear operators, it may or may not be true for a non-linear map. (To prove boundedness$\Rightarrow$continuity, write

$$L(f_k,f_k) - L(f,f) = \int B(\theta)[f_k'(f_{1,k}'-f_1') + f_1'(f_k'-f')]d\theta d\varepsilon d^3v_1,$$

and integrate over d^3v).

<u>PROOF OF THEOREM I</u>. We first show the sequence $\{n_k\}$ is monotone increasing ($\vec{v}$-a.e.). Clearly $n_1(t) \geqslant n_0(t) = 0$. By induction, if $n_k(t) \geqslant n_{k-1}(t)$, $L_k(t) \geqslant L_{k-1}(t)$ (Eq. (54b))$\Rightarrow n_{k+1}(t) \geqslant n_k(t)$. Furthermore, $n_k(t) \in L^1(R^3) \Rightarrow n_{k+1}(t) \in L^1(R^3)$, by Eqs. (60), so

$$\rho_k(t) \equiv \int d^3v n_k(t) \tag{65}$$

exists. Furthermore since $\{n_k\}$ is a positive monotonic sequence so is $\{\rho_k(t)\}$. We now show that the $\rho_k(t)$ are uniformly bounded by ρ. Integrate Eq. (63) to obtain

$$\rho_{k+1}(t) = \rho e^{-\nu t} + V\sigma\int_0^t e^{-\nu(t-\tau)}\rho_k^2(\tau)d\tau. \tag{66}$$

Again, we use an inductive argument. Note $\rho_0(t) = 0 \leqslant \rho$. Assume $\rho_n(t) \leqslant \rho$. Then

$$\rho_{n+1} \leqslant \rho e^{-\nu t} + V\sigma\rho^2\int_0^t e^{-\nu(t-\tau)}d\tau \tag{67a}$$

$$= \rho e^{-\nu t} + \frac{V\sigma\rho^2}{\nu}(1-e^{-\nu t}) = \rho \tag{67b}$$

recalling that $\nu = V\sigma\rho$. Thus, we have shown that $\{n_k\}$ is a monotonically increasing sequence of integrable (positive) functions on R^3 and that $\{\rho_k\} = \{\int d^3vn_k\}$ is monotonic and uniformly bounded in t. Thus $\lim_{k\to\infty}\rho_k(t)$ exists. It follows, from the Lebesgue monotone convergence theorem[10] that $\lim_{k\to\infty} n_k(t) \equiv n(t)$ exists. Furthermore, since L is continuous, $L_k = L(n_k,n_k)\to L(n,n)$ so n obeys Eq.(63). This proves the theorem.

The question still remains as to whether or not $n(\vec{v},t)$ is a solution of Eq. (62), i.e. a strong solution, since Theorem I proves only the existence of a mild solution. It is rather easy to show that n(t) is differentiable in $L^1(R^3)$, i.e. $\lim_{h\to 0}\frac{1}{h}\|n(t+h)-n(t)\|_{L^1}$ exists, so n is a Banach-space solution of the original equation. (Additional assumptions are necessary to prove that a <u>classical solution</u> exists, i.e. that $u'(t)$ exists pointwise. We do not consider that question further).

<u>COROLLARY I</u>. If n(t) is a solution to Eq. (63), then the $\lim_{h\to 0}\frac{1}{h}\|n(t+h) - n(t)\|$ exists $0 \leqslant t \leqslant T < \infty$, where $\|\cdot\|$ refers to the L^1 norm.

<u>PROOF</u>. From Eq. (63) we easily compute

$$\lim_{h\to 0}\|n(t+h) - n(t)\| \leqslant \|f_0\|(\nu e^{-\nu t}) + \frac{d}{dt}\int_0^t \|L(n,n)\|e^{-\nu(t-\tau)}d\tau + \lim_{h\to 0}\frac{1}{h}\int_t^{t+h}\|L(n,n)\| \quad e^{-\nu(t+h-\tau)}d\tau \tag{68a}$$

$$\leqslant \|f_0\| + 2\sup_{0<t<T}\|L(n,n)\|. \tag{68b}$$

6. <u>EXISTENCE AND UNIQUENESS, HARD SPHERES.</u>

Historically, the first attempt to treat the question of existence of solutions to the Boltzmann equation was carried out by T. Carleman,[11] whose model was a gas of hard spheres (in the spatially homogeneous situation). This problem is considerably more difficult than that of pseudo-Maxwell molecules, described in Sec. 5, and involves rather lengthy computations, so we shall content ourselves with sketching the techniques.

We recall that in the previous section there was no restriction put on the initial datum, $f_0(\vec{v})$. In fact, f_0 could be any function in $L^1(R^3)$. For the hard-sphere gas and, in fact, for other more general scattering models,[12] existence can still be proved if the initial datum is sufficiently restricted. The equation Carleman considered was (cf. Eq. (54a))

$$\frac{\partial n(\vec{v},t)}{\partial t} = r_0^2 \int d\theta d\varepsilon d^3v_1 V \sin\theta \cos\theta (n' n_1' - nn_1) \tag{69a}$$

$$= L(n,n) - \nu(n)n. \tag{69b}$$

(The notation here differs from that of Carleman; it was chosen to be consistent with that of Grad.[5]) We should point out that ν was a constant independent of n for the pseudo-Maxwell gas considered in Sec. 5. That is not the case for the gas of hard sphere we are considering now.

Crucial to Carleman's analysis is the following <u>Lemma</u>. If one has, for $0 \leqslant t \leqslant t_1$,

$$\frac{du}{dt} + p(t)u \leqslant q(t), \quad p(t) > 0, \tag{70}$$

one will have, u_0 designating the initial value $u(0)$

$$\sup_t u(t) \leqslant \max\left\{u_0, \sup_t \frac{q(t)}{p(t)}\right\}. \tag{71}$$

Now, in order to demonstrate the existence of a solution of Eq. (69) which for $t=0$ reduces to a given function $f_0(\vec{v})$, Carleman applies a method of successive approximations which, at the same time, permits one to conclude that the solution is positive.

Suppose that the function $f_0(\vec{v})$ is continuous and that it satisfies the inequalities

$$0 \leqslant f_0(\vec{v}) \leqslant \frac{a}{(\sqrt{1+v^2})^\kappa}, \quad \kappa > 6, \tag{72}$$

for some constant a. Define recursively a sequence of functions $\{n_k\}$ by

$$n_0(\vec{v}) = f_0(\vec{v})e^{\gamma t},\ \gamma > 0 \tag{73a}$$

and

$$\frac{\partial n_k}{\partial t} + \nu(n_{k-1})n_{k-1} = L(n_{k-1},n_{k-1}), \tag{73b}$$

with

$$[n_k(\vec{v},t)]_{t=0} = f_0(\vec{v}). \tag{73c}$$

One sees immediately that all the n_k are positive, since f_0 is positive.

Carleman goes on to show that it is possible to choose the constant and the interval $0<t<t_0$ so that the functions n_k are uniformly bounded (in k) on $[0,t_0]$. In proving this, Eq. (70) is used as well as the inquality for n obeying (72):

$$L(n,n) < \frac{8\pi a\rho}{(\kappa-2)(\sqrt{1+v2})^{\kappa-1}} + \frac{K(\kappa)}{(\sqrt{1+v^2})^{\kappa}} = S(n), \tag{74}$$

where, as before, $\rho = \int n(\vec{v})d^3v$, and K is a constant. Lengthy computations lead to the result that $\exists\tau$ s.t.

$$n_k \leqslant \frac{ae^{\gamma t}}{(\sqrt{1+v^2})^{\kappa}}\ ,\ 0 < t < \tau \tag{75}$$

Carleman then goes on, again after lengthy computations, to show that the sum

$$\Sigma(n_k(\vec{v},t) - n_{k-1}(\vec{v},t)) \tag{76}$$

is dominated by

$$\frac{\Sigma Ce^{\beta t}(1-\frac{\alpha}{4})^n}{(\sqrt{1+v^2})^{\kappa}} \tag{77}$$

where β and C are constants. One concludes that $n_k(\vec{v},t)$ converges uniformly in $0<t<\delta<\tau$ to a limit $n(\vec{v},t)$. He goes on to show that $L(n_k,n_k)\to L(n,n)$ and $\nu(n_k)\to\nu(n)$. This proves existence on the interval $0<t<\delta$, but one can take $n(\delta)$ as the initial value for the next time interval and, since it still obeys (72), extend the solution to 2δ and, by iteration, to $n\delta$. He goes on to prove uniquesness in the standard way by assuming the existence of two solutions,

$n^{(1)}$ and $n^{(2)}$, say, and showing that $n^{(1)} - n^{(2)}$ is identically zero, again utilizing the various a priori bounds used in the existence proof. His final results is stated as

THEOREM II. Let a continuous function $f_0(\vec{v})$ be given, satisfying the condition

$$0 \leqslant f_0(\vec{v}) \leqslant \frac{a}{(\sqrt{1+v^2})^{\kappa}}\ , \quad \kappa < 6.$$

Then there exists a continuous solution $n(\vec{v},t)$ of Eq. (69), reducing for $t=0$ to $f_0(\vec{v})$. $n(\vec{v},t)$ satifies the inequalities

$$0 \leqslant n(\vec{v},t) < \frac{C}{(\sqrt{1+v^2})^{\kappa}}$$

where C is a constant independent of t and of $\vec{v}$. There does not exist another solution taking the same initial value and satisfying in an interval $0 \leqslant t < t_1$ an inequality of the form

$$|n| < \frac{C}{(\sqrt{1+v^2})^{\kappa}}, \quad \kappa > 6, \ C \text{ a constant.}$$

7. OTHER GLOBAL EXISTENCE PROOFS.

The existence proof presented in Sec. 5, as given by H. Grad,[5] was based on the earlier work of E. Wild[12] and D. Morgenstern[13] and, except for minor details, is essentially equivalent to their treatment. Since the appearance of Grad's article in 1968, more attention has been focused on the space-dependent equation--discussed in Chapters III-VI--but one should at least call attention to the work of Arkeryd on the spatially homogeneous Boltzmann equation.[14]

Arkeryd's first result is to prove the existence and uniqueness of a solution to the Boltzmann equation written in the form

$$\frac{df}{dt} + hf = \overline{G}f, \quad t > 0, \quad f(0) = f_0 \geqslant 0 \tag{78}$$

where h and $\overline{G}$ are both bounded. This is, then, a generalization of the proof presented in Sec. 5, where a specific collision model, i.e. Maxwell molecules, was chosen. His technique is essentially the same as that given in Sec. 5, that is the construction of the Wild-Morgenstern monotonic sequence of iterates. However, he proves more. His most important result is to derive a rigorous H-theorem for this case. (Recall that we pointed out in Chapter I that the "derivation" given there of the H-theorem was only heuristic.)

PROPOSITION. Let $f(\vec{v},t)$ be a solution of Eq. (78). Then if $f_0 \ln f_0$ and $f_0(\vec{v})(1+v^2) \in L^1(R^3)$, then $f(\vec{v},t)\ln f(\vec{v},t) \in L^1(R^3)$ and $H(t) = \int f(\vec{v},t)\ln f(\vec{v},t)d^3v$ is a non-increasing function of t.

The proof is long and technical, and depends in an essential way on Fatou's Lemma and the Lebesgue dominated convergence theorem. We refer the reader to Reference 14(a) for details.

Another result of significance proved by Arkeryd is the existence of a solution to a Boltzmann equation for which the operators h and $\bar{G}$, Eq. (78), are not separately bounded. His proof depends crucially on the following compactness result:

Lemma. Let $(f_n)_1^\infty$ be a sequence of non-negative $L^1(R^3)$-functions such that $\int v^{2\kappa}|f_n|d^3v \leq C_\kappa$, n=1,2,... for some $\kappa > 0$ and such that

$$\int_{R^3} f_n(\vec{v})\ln f_n(\vec{v})d^3v \leq C$$

where C_κ and C are constants. Then $(f_n)_1^\infty$ contains a subsequence $(f_{n_j})_{j=1}^\infty$ converging weakly to a function $f \in L^1$ and for any collision invariant $\phi(\vec{v})$

$$\lim_{j\to\infty} \int f_{n_j}(\vec{v})\phi(\vec{v})d^3v = \int f(\vec{v})\phi(\vec{v})d^3v$$

if

$$(1+|v|)^{-\kappa'}\phi(\vec{v}) \in L^\infty(R^3),\ 0 < \kappa' < \kappa.$$

Thus, for non-cutoff potentials, a solution exists as a weak limit of solutions to a sequence of cutoff potentials. We shall meet this compactness argument again in Chapter VI, where the Boltzmann equation is treated on a spatial lattice.

References 14(b) and (c) are mostly concerned with positivity, asymptotic behavior, and existence of higher moments of $f(\vec{v})$. Again, the reader is referred to the references for details.

The papers of Povzner[15] and of Cercignani et al[36] might also be mentioned. These are both concerned with the spatially dependent equation. In Reference 15, a "mollifier" is introduced (cf. Chapter IV) and the claim is made that the spatially homogeneous result is contained therein as a special case. Reference 16 deals with the Boltzmann equation on a lattice (cf. Chapter VI). The spatially homogeneous case is certainly contained as a special case there--one simply considers a lattice consisting of a single point.

Finally, existence of global solutions to the spatially homogeneous Boltzmann equation with non-Maxwellian molecules was also obtained by Di Blasio[16] using entirely different methods. In particular, she modified the non-linear semigroup theory of Crandall and Liggett so that it included the model of the Boltzmann equation, including also external force terms. Di Blasio was able also to obtain new results on the rate of convergence to the equilibrium solution.

"...the complications of existence."

Henry James, The Portrait of a Lady

III. LOCAL EXISTENCE THEORY FOR THE SPACE-DEPENDENT EQUATION

8. BASIC DIFFICULTIES WITH THE SPACE-DEPENDENT PROBLEM.

It seems at first glance that the procedure carried out in Chapter II for proving existence for the spatially homogeneous Boltzmann equation could be carried over to the spatially inhomogeneous case without any particular difficulty. In fact, the free propagator (to adopt the terminology of quantum field theory) is replaced by $n(r,v,t) = e^{-\int_0^t \nu(\vec{r}-\vec{v}s)ds}\, n(\vec{r}-\vec{v}t,\vec{v},o)$. This well-known result can be obtained either by integrating the equation

$$\frac{\partial n}{\partial t} + \vec{v}\cdot\vec{\nabla}n + \nu n = 0 \tag{79}$$

or by reasoning physically that the particle density at position $\vec{r}$, traveling with velocity $\vec{v}$ at time t is exactly that at position $\vec{r}-\vec{v}t$ at time zero traveling in direction $\vec{v}$, diminished exponentially by collisions which remove molecules from the direction $\vec{v}$. We see the missing term in Eq. (79) is simply the "creation" term identified in Chapter I, when the Boltzmann equation was derived. Strictly speaking, Eq. (79) should be called the "destructive" rather than "free" equation. We could then identify the free propagator U_t by

$$(U_t f)(\vec{r},\vec{v},t) = f(\vec{r}-\vec{v}t,0) \tag{80a}$$

or the "destructive" propagator

$$(D_t f)(\vec{r},\vec{v},t) = e^{-\int_0^t \nu(\vec{r}-\vec{v}s)ds}\, f(\vec{r}-\vec{v}t,\vec{v},o). \tag{80b}$$

It is a triviality to demonstrate that U_t is a unitary group and D_t a contraction semi-group. The realization (80b) of the semi-group D_t, incidentally, is based on the pseudo-Maxwell collision model, so that $\nu = \sigma\rho$, independent of $\vec{V}$ (cf. Sec. 5). For more general collision models, D_t is a non-linear semi-group (i.e., ν depends on n) but can still be shown to be a contraction as, indeed, physical reasoning tells us it must be.

Thus, it would appear that one could use either U_t or D_t to generate an integral equation from the Boltzmann equation, and proceed to prove existence for the spatially dependent equation in direct analogy with the proof presented for the spatially homogeneous equation. However, a very basic difficulty arises which has never been fully overcome. The natural function space in which to seek solutions to the Boltzmann equation is L^1, both because the quantity $n(\vec{r},\vec{v},t)$ is a density and because the conservation laws, which involve integration over space and velocity variables, are of considerable help in proving existence(recall, in Sec. 6 that we used the conservation law $\frac{d\rho}{dt} = 0$ from the beginning). However, the collision integral $J(n,n)$ is not in general integrable with respect to d^3r since it involves the product of two L^1 functions nn_1. This non-integrability does not arise when integration is carried out with respect to d^3v because the cross section "mollifies" the velocity dependence, i.e. n and n_1 involve two different velocities, $\vec{v}$ and $\vec{v}_1$ or $\vec{v}'$ and $\vec{v}_1'$ with a kernel $\sigma(|v-v_1|)$ which has the property that $\int nn_1\sigma d^3v_1$ and $\int n'n_1'\sigma d^3v_1$ are both L^1 functions of $\vec{v}$. Since the spatial dependence of n is not mollified--n and n_1 are both evaluated at the same spatial point $\vec{r}$--the L^1 norm is not appropriate for dealing with the spatially homogeneous equation, and so a sup norm must generally be used, i.e. solutions are sought in L^∞, the space of (essentially) bounded functions. In such a space, it is generally possible to prove existence only for sufficiently short time--so-called "local existence." Attempts to move on to subsequent time intervals, using the solution to the first intervals as the datum for the second, and so forth, have not in general been successful because it has not been possible to show that the solution remains adequately bounded. (Another technique, described in Chapter IV, is to introduce a spatial mollifier, and then to work in L^1. Since a mollifier changes the basic structure of the Boltzmann equation, in effect introducing a non-local interaction between molecules, it would be desirable to prove the existence of a limit as the mollifier approaches a delta function. So far, this has not been accomplished.)

A number of local existence proofs have been presented. We shall reproduce here the first, that presented originally by H. Grad,[5] which has the advantage of simplicity but the disadvantage of being restricted to pseudo-Maxwell molecule. We shall then mention how other authors have generalized Grad's result to some extent, without going into a great deal of detail.

9. GRAD'S PROOF OF LOCAL EXISTENCE: PSEUDO-MAXWELL MOLECULES.

Introduce the abbreviations

$$[n] = n'n_1' - nn_1$$

and

$$\{n\} = n'n_1' + nn_1.$$

Then

$$J(n,n) = \int B(\theta)[n]d\theta d\varepsilon d^3v_1 \tag{81}$$

and we recall that $v\sigma$ is a finite constant, independent of v:

$$v\sigma = \int B(\theta)d\theta d\varepsilon = \hat{\sigma}, \text{ a constant.}$$

We shall frequently write $J(n,n) = J(\vec{r},\vec{v},t)$ or, simply, J. Assume that $n(\vec{r},\vec{v},0) = f_0$ is uniformly bounded by a Maxwellian:

$$f_0 \leq M(v). \tag{82}$$

We use the free propagator U_t to convert Eq. (12) to an integral equation:

$$n(\vec{r},\vec{v},t) = U_t f_0(\vec{v}) + \int_0^t U_{t-\tau} J(n,n)(\tau)d\tau \tag{83a}$$

$$= f_0(\vec{r}-\vec{v}t,\vec{v}) + \int_0^t J(\vec{r}-\vec{v}(t-\tau),\vec{v},t)d\tau. \tag{83b}$$

The modus operandi will be to construct an iterative solution to Eq. (83) and to prove convergence--for t sufficiently small. We shall work in the L^∞ space with norm

$$\|f\| = \sup_{\vec{r},\vec{v},t} \left(\frac{|f|}{M}\right) \tag{84}$$

where M is the Maxwellian bound introduced in Eq. (82).

The following Lemma is crucial to the convergence proof:

LEMMA 1. Let $0 \leq n(\vec{r},\vec{v},t) \leq 2M(v)$. Then $|J| \leq 8\hat{\sigma}RM(v)$ where $R = \int M(v)d^3v$.

PROOF. Since for $n \geq 0$, $\{n\} \geq [n]$,

$$|J| \leq \int B(\theta)\{n\}d\theta d\varepsilon d^3v_1$$

$$\leq 4\int B(\theta)\{M\}d\theta d\varepsilon d^3v_1,$$

since $n \leq 2M$. But the Maxwellian, being an equilibrium distribution, obeys

[Eq. (14a)]

$$M(v)M(v_1) = M(v')M(v_1').$$

Thus

$$|J| \leq 8\int B(\theta)M(v)M(v_1)d\theta d\varepsilon d^3v_1 = 8\hat{\sigma}RM(v),$$

proving the Lemma.

Now consider the time interval $0 \leq t \leq T = \frac{1}{8\hat{\sigma}R}$ and establish an iteration scheme for Eq. (83) as follows:

$$n_{k+1} = U_t f_0 + \int_0^t U_{t-\tau} J_k(\tau)d\tau, \tag{85a}$$

$$J_k = \int B(\theta)[n_k]d\theta d\varepsilon d^3v_1 \tag{85b}$$

with the initial iterate taken to be zero. Then we have

LEMMA 2. For $0 \leq t \leq T$, if f_0 obeys Eq. (82), then for the iteration scheme defined by Eqs. (85), $0 \leq n_k \leq 2M(v)$.

PROOF. We proceed by induction. First, $0 \leq n_1 \leq f_0 \leq 2M(v)$ by hypothesis. Assume now that $n_k \leq 2M(v)$. Then from Eq. (85a)

$$n_{k+1} \leq U_t f_0 + \int_0^t |U_{t-\tau} J_k(\tau)|d\tau.$$

But since f_0 and J_k are both uniformly bounded,

$$n_{k+1} \leq M(v) + \int_0^t |J_k(\tau)|d\tau \leq M(v)(1+8\hat{\sigma}Rt) \leq 2M(v)$$

since $t < \frac{1}{8}\hat{\sigma}R$. Furthermore, n_{k+1} is non-negative since for $t < \frac{1}{8}\hat{\sigma}R$, $U_t f_0$ dominates $\int_0^t U_{t-\tau}J_k(\tau)d\tau$ by Lemma 1. This proves the Lemma.

We now state

LEMMA 3. The sequence (n_k) defined by the iteration scheme (85) converges in the L^∞-norm [Eq. (84)] for $t < T$

PROOF. Compute

$$n_k - n_j = \int B([n_k] - [n_j])d\theta d\varepsilon d^3v_1.$$

Simple algebraic manipulation using Eq. (80a) yields the inequality

$$[n_k] - [n_j] \leq n_k'|n_{1,k}'-n_{1,j}'| + n_{1,j}'|n_k'-n_j'| + n_k|n_{1,k}-n_{1,j}| + n_{1,j}|n_k-n_j|.$$

"De non apparentibus et de non existentibus non ratio."

Maxim

IV. THE MOLLIFIED BOLTZMANN EQUATION

11. GENERAL CONSIDERATIONS.

We have seen in Chapter III how the product of the densities $n(\vec{r},\vec{v},t) \times n(\vec{r},\vec{v},t)$ evaluated at the same spatial point $\vec{r}$ makes it impossible to seek solutions to the Boltzmann equation in the space L^1. The necessity of dealing instead with L^∞ restricts existence proofs to local existence, i.e. sufficiently small t, since it is in general not possible to prove that the requisite bounds to the solution persist in time.

This difficulty did not arise in the space-homogeneous case even though the product $n(\vec{v})n(\vec{v}_1)$ occurred in the collision term. The point is that in the map $J: L^1(R^3) \times L^1(R^3) \to L^1(R^3)$ the kernel $B(\theta,V)$ acts as a mollifier so that the L^1-norm of $J(n,n)$ exists.

This suggests the introduction of a spatial mollifier into the Boltzmann equation, so that the collision integral is written

$$J_h(n,n) = \int B(\theta,V)h(\vec{r},\vec{r}_1)[n(\vec{r},\vec{v}',t)n(\vec{r}_1,\vec{v}_1',t)-n(\vec{r},\vec{v},t)n(\vec{r}_1,\vec{v}_1,t)]d\theta d\varepsilon d^3v_1 d^3r_1.$$

If the mollifier $h(\vec{r},\vec{r}_1)$ is chosen appropriately, then J will be an element of $L^1(R^3 \times R^3)$, and a procedure analogous to that used for the spatially homogeneous equation in Chapter II will be successful for proving existence. The idea of the mollifier was introduced by Morgenstern,[19] and later extended by Povzner.[15] We shall give Morgenstern's proof in some detail, and sketch the ideas put forth by Povzner.

Before proceeding with the mathematical details, we should make a few comments. First, a similar difficulty arises in the theory of introducing quantum fields with popular interactions like $\lambda\phi^4$.[20] Since the field ϕ is supposed to be an operator-valued distribution, powers like ϕ^4 are not well-defined. (Even if one deals with the smeared fields, i.e. $\phi(f)$ for f a test function, the difficulty is still present in that $\phi(f)$ is supposed to be L^2, not L^4.) The introduction of a mollifier allows computations to be carried

Dividing by $M(v)M(v_1)$ and using Eq. (84) yields (recalling $M(v)M(v_1) = M(v')M(v_1'))$

$$\frac{|[n_k]-[n_j]|}{M(v)M(v_1)} \leqslant 2\|n_{1,k}'-n_{1,j}'\| + 2\|n_k'-n_j'\| + 2\|n_{1,k}-n_{1,j}\| + 2\|n_k-n_j\|,$$

where we have used Lemma 2. Since clearly $\|f\| = \|f'\| = \|f_1\| = \|f_1'\|$ this gives

$$\frac{|[n_k]-[n_j]|}{M(v)M(v_1)} \leqslant 8\|n_k-n_j\|.$$

Thus from Eq. (85b)

$$|J_k-J_j| \leqslant \int B(\theta)|[n_k]-[n_j]|d\theta d\varepsilon d^3v_1 \leqslant 8\|n_k-n_j\| \int B(\theta)M(v)M(v_1)d\theta d\varepsilon d^3v_1$$

$$= 8\hat{\sigma}R\|n_k-n_j\|M(v)$$

from which

$$\|J_k-J_j\| \leqslant 8\hat{\sigma}R\|n_k-n_j\| .$$

Using (85a) plus the above estimate, we readily compute

$$\|n_k-n_j\| \leqslant 8\hat{\sigma}Rt\|n_{k-1}-n_{j-1}\| .$$

For $t < \frac{1}{8}\hat{\sigma}R$ we conclude that the sequence (n_k) is Cauchy, and hence convergent, in the L^∞ norm. This proves Lemma 3. Call the limit function n. In the process of proving Lemma 3 we have also proved:

COROLLARY. The sequence (J_k) converges, in L^∞, to a limit J.

Furthermore, n and J evidently obey the equation

$$n(\vec{r},\vec{v},t) = U_t f_0 + \int_0^t U_{t-\tau} J(\tau)d\tau, \tag{86}$$

that is, $n(\vec{r},\vec{v},t)$ is a mild solution of the Boltzmann equation. (Sometimes, as we shall see in later Chapters, compactness arguments are used to deduce the convergence, in which case it is usually not possible to conclude that the limit function is even a mild solution of the original equation.)

To prove that $n(\vec{r},\vec{v},t)$ is actually a strong solution of the Boltzmann equation we need

LEMMA 4. Let $f_0(\vec{r},\vec{v})$ be continuous in $\vec{r}$ and $\vec{v}$. Then $\frac{\partial n}{\partial t} + \vec{v}\cdot\vec{\nabla}n$ exists, so that n is a strong solution of the Boltzmann equation.

PROOF. We observe from Eq. (86) that $\frac{\partial U_t f_0}{\partial t} = -\vec{v}\cdot\vec{\nabla}U_t f_0$, since U_t is the free propagator. That is, $\frac{\partial}{\partial t} + \vec{v}\cdot\vec{\nabla}$ exists for continuous f_0 even though f_0

may not be differentiable in either t or $\vec{r}$. From the iteration scheme (85) we see that $(\frac{\partial}{\partial t} + \vec{v}\cdot\vec{\nabla})\, n_k$ exists for every k. Since $n_k \to n$ uniformly in $\vec{r}$ and t (for $t < T$), $(\frac{\partial}{\partial t} + \vec{v}\cdot\vec{\nabla})\, n$ also exists.

We can restate the above results as a Theorem:

THEOREM III. Let $0 \leqslant f_0(\vec{r},\vec{v})$ be the initial datum for the Boltzmann equation, such that $f_0(\vec{r},\vec{v}) \leqslant M(v)$ for some Maxwellian distribution M. Then for $t \leqslant \frac{1}{8}\sigma R$, $R = \int M(v)d^3v$, there exists a mild solution to the Boltzmann equation which is everywhere non-negative. If f_0 is continuous in $\vec{r},\vec{v}$ then the mild solution is also a strong solution. (This theorem applies, of course, to a pseudo-Maxwell molecule scattering model.)

It is important to remember that the time and space derivatives of the solution need not exist separately.

10. GENERALIZATION OF GRAD'S RESULT.

We observe that the question of boundary conditions never entered the analysis presented in the previous section. In fact, boundary effects are hidden in the free propagator U_t. That is, $(U_t f)(\vec{r},\vec{v},t) = f(\vec{r}-\vec{v}t,\vec{v},0)$ for $t < \frac{R}{v}$ where R is the distance from $\vec{r}$ to the boundary in the direction $-\vec{v}$. For $t > \frac{R}{v}$ the definition of U_t must be modified to account for the reflection. For example, for specular reflection, $U_t f$ would be defined for $t > \frac{R}{v}$ by $(U_t f)(\vec{r},\vec{v},t) = f(\vec{r}_s - \vec{v}'(t-\frac{R}{v}),0)$ where $\vec{v}'$ is related to the components of $\vec{v} = (\vec{v}_{\parallel}, v_{\perp})$ by $\vec{v}' = (\vec{v}_{\parallel}, -v_{\perp})$ and $\vec{r}_s$ is the point on the surface along the line $-\vec{v}$ from $\vec{r}$. These ideas are made more precise by Kaniel and Shinbrot,[17] who also relax Grad's assumption of pseudo-Maxwell molecules to some extent. As did Grad, these authors worked with a sup norm, and assumed the initial data to be Maxwellian-bounded. Their basic technique is to consider a linear problem associated with the Boltzmann equation

$$\frac{\partial n}{\partial t} + \vec{v}\cdot\vec{\nabla}n = J(n,\hat{n}) \tag{87}$$

where $\hat{n}$ is a given function. Then, if $\hat{n}$ is known, n can be found by standard techniques for treating linear equations. To deal with the non-linear equation recursively, assuming an initial datum f_0, two sequences are generated, (1_k)

and (n_k) with $u_0 = f_0$, l_0 = the solution of Eq. (86) with $\hat{n} = u_0$, u_1 the solution of (86) with $\hat{n} = l_0$ and so forth. They are able to prove that for sufficiently small T the sequences (n_k) and (l_k) obey $0 \leqslant l_0(t) \leqslant l_1(t) \leqslant \cdots \leqslant l_n(t) \leqslant u_k(t) \leqslant \cdots \leqslant u_0(t)$, $0 \leqslant t \leqslant T$. That is, the sequence (l_k) (resp. (u_k)) is a lower bound (resp. upper bound) to the true solution and because they are bounded and monotone, they converge.

This technique, of approximating the non-linear equation by a sequence of linear equations, is one which we shall meet again (in Chapter VI).

Two other papers worthy of mention have been written by A. Glikson.[18] Glikson works in the same function space as Grad, but generalizes to non-Maxwellian molecules (although, as always, retaining the cutoff assumption) and allows an external force field to be present. He then obtains basically the same result as Grad using, however, a fixed point-contraction mapping principal.

The second of Glikson's two papers is couched in somewhat more abstract terms; the results are basically the same as in the first paper, except some additional information is available--existence of moments, continuous dependence of the solution on the initial data, etc.

out (for example, the calculation of correlation functions--Schwinger or Wightman functions) after which the limit as the mollifier approaches a delta function can be taken on the correlation functions. Such a limit has not been proved for the mollified Boltzmann equation, one difficulty being that one generally deals with the density itself, rather than with correlation functions since, for the Boltzmann equation there is no analogue of the vacuum expectation value used to define the Schwinger or Wightman functions in quantum field theory.

Because of the absence of a proof of a delta function limit, the mollified equation is defended in both References 19 and 15 (especially in Reference 19) as having more physical reality than the unmollified version of the equation. Certainly the difficulties which have been encountered in proving existence for the spatially inhomogeneous Boltzmann equation lead the casual observer to suspect the possible truth of this conjecture.

Another approach to the spatially inhomogeneous problem, also inspired by quantum field theory, is to set the equation on a spatial lattice. That is the subject of Chapter VI of these notes.

12. MORGENSTERN'S EXISTENCE PROOF.

Morgenstern first considers an abstract bilinear evolution equation, and then applies the result to the Boltzmann equation.

Consider then a measure space Ω with points ω, the real Banach space $L^1(\Omega)$ with norm $\|f\| = \int |f| d\omega$ and a bounded bilinear form $A(t)$: $L^1 \times L^1 \to L^1$ strongly continuous in t; and let

$$\|A\| = \sup_{f,g \in L^1(\Omega)} \frac{\|A(f,g)\|}{\|f\| \, \|g\|} = a. \tag{89}$$

Take $t_0 < \frac{1}{4a}$ and consider solutions to the equation

$$\frac{\partial f(t,\omega)}{\partial t} = A(t;f(t),f(t))(\omega), \quad f(0) = f_0, \quad \|f_0\| \leq 1. \tag{90}$$

Then

LEMMA 1. For $0 \leq t \leq t_0$, there exists a unique, strongly continuous solution to Eq. (90), $f(t,\cdot) \in L^1(\Omega)$. This solution, which depends continuously on f_0 may be obtained by the iterative scheme $f_0(t) = f_0$; $f_{n+1}(t) = f_0 + \int_0^t A(\tau; f_n(\tau), f_n(\tau))d\tau$.

PROOF. The operator $\Phi(F) \equiv f_0 + \int_0^t A(\tau ; F(\tau), F(\tau))d\tau$ is easily seen to map the ball $B = \{f(t,\cdot)\varepsilon L^1(\Omega) \mid \|f(t)\| \leqslant 2,\ t < t_0\}$ into itself. Furthermore, a simple computation shows that Φ is a contraction on B. The result follows from the contraction mapping principle.[21]

Next, suppose that A can be decomposed into the difference of two non-negative forms so that Eq. (90) becomes

$$\frac{\partial f(t,\omega)}{\partial t} = -f(t,\omega)R(t;f(t))(\omega) + M(t;f(t),f(t))(\omega),\ f(0,\omega) = f_0(\omega) \geqslant 0,$$

$$\|f_0\| \leqslant 1, \tag{91}$$

where R: $L^1 \to L^\infty$ is non-negative, $\|R\| \leqslant m < \infty$ and R is continuous in t as a map from the L^1-strong to the L^∞ weak star topology. (That is, $\forall \varepsilon > 0$, $\exists \delta$ s.t. $\forall f,g \ \varepsilon\ L^1$, $\int |g(R(t,f) - R(t_0,f))| < \varepsilon$ for $|t_1 - t_0| < \delta$.)

M: $L^1 \times L^1 \to L^1$, $\|M\| \leqslant m$, M strongly continuous in t. (As we shall see, this representation corresponds to decomposing J into L-ν as in Chapter II. This scheme has the advantage that the solutions corresponding to positive initial data can be shown to be positive.) Morgenstern proves

LEMMA 2. Define $\hat{t}_0$ by $5\hat{t}_0 m + 4\hat{t}_0^2 m^2 = 1$. Then for $t < \hat{t}_0$ Eq. (91) has a unique, non-negative continuous solution $f(t,\cdot)\varepsilon L^1$, depending continuously on f_0 which agrees with the solution of Eq. (90) for $0 \leqslant t < \min(t_0, \hat{t}_0)$.

PROOF. Use the iteration scheme

$$\frac{\partial f_{n+1}}{\partial t} = -f_{n+1}R(t;f_n) + M(t;f_n,f_n) \tag{92}$$

and convert to an integral equation for f_{n+1} in the usual way, i.e.

$$f_{n+1} = f_0(\omega)e^{-\int_0^t R(\tau;f(\tau))d\tau} + \int_0^t M(\tau;f_n(\tau),f_n(\tau))e^{-\int_\tau^t R(\sigma;f_n(\sigma))d\sigma}\,d\tau. \tag{93}$$

If we rewrite this as

$$f_{n+1}(t) = G(f_n) \tag{94}$$

then more or less straightforward computations show (i) G maps $\hat{B} = \{f(t,\cdot)\varepsilon L^1(\Omega) \mid \|f(t)\| \leqslant 2,\ t \leqslant \hat{t}_0\}$ into itself and (ii) G is a contraction on $\hat{B}$. Thus a solution exists. Positivity is evident, and uniqueness implies the solutions of Eqs. (90) and (91) are the same if we identify $A = -R + M$.[22] Finally

<u>THEOREM IV</u>. Suppose $\int d\omega A(t;f(t,\omega),f(t,\omega)) = \int d\omega[-f(t,\omega)R(t;f(t,\omega)) + M(t;f(t,\omega),f(t,\omega))] = 0\ \forall f \varepsilon L^1(\Omega)$. Then for all t the initial value problem (92) has a unique non-negative solution, depending continuously on the initial value.

<u>PROOF</u>. By Lemma 2, there is a non-negative solution for $0 \leqslant t \leqslant t_0$, where t_0 depends only on $\|f_0\|$. But integrating Eq. (92) gives $\int d\omega \frac{\partial f(t,\omega)}{\partial t} = \frac{d}{dt}\|f(t,\omega)\| = 0$, so t_0 can be chosen as the starting point for the next time step, and the solution exists for $t_0 \leqslant t \leqslant 2t_0$. Since this procedure can be carried out ad infinitum, the theorem follows.

Observe the important role played by positivity in the proof of this theorem.

To apply this result to the mollified Boltzmann equation

$$\frac{\partial n(\vec{r},\vec{v},t)}{\partial t} + \vec{v}\cdot\vec{\nabla}n = J_h(n,n), n(\vec{r},\vec{v},0) = f_0(\vec{r},\vec{v}) \geqslant 0 \tag{95}$$

(with J_h given by Eq. (88)) it is necessary first to rewrite Eq. (95) in the form of Eqs. (90), (91), i.e. to dispense with the $\vec{v}\cdot\vec{\nabla}$ term. Since $\vec{v}\cdot\vec{\nabla}$ is the generator of the unitary group U_t introduced in Chapter III [Eq. (80a)], this can be done by noting that the Boltzmann equation is equivalent to

$$\frac{\partial}{\partial t}(U_{-t}n)(\vec{r},\vec{v},t) = U_{-t}J(n,n) = J(U_{-t}n, U_{-t}n) \tag{96}$$

and to define $U_{-t}n = f(\vec{r},\vec{v},t)$. Note that the boundary conditions are hidden in the definition of U_{-t}, as we have already pointed out in Sec. 10. It is necessary, however, that the boundary condition, whether it be diffuse or specular reflection or anything else be conservative. (For example, if molecules are created by collisions with the walls, no finite solution can exist, while if they are destroyed, all solutions will tend to zero.)

We take the point of view that the mollifier is introduced after the change of variable $n \to f$ is made. We thus are considering the equation

$$\frac{\partial f}{\partial t} = J_h(f,f) \tag{97a}$$

$$= -\nu_h(f)f + L_h(f,f). \tag{97b}$$

The mollifier h is assumed to have the following properties:

(i) $h(\vec{r},\vec{r}_1) = h(\vec{r}_1,\vec{r}) \quad 0, \; h(\vec{r},\vec{r}) > 0$

(ii) $\sup\limits_{\vec{r},\vec{r}_1} h(\vec{r},\vec{r}_1) = h < \infty$.

We might also require that h be normalized:

(iii) $\int d^3r h(\vec{r},\vec{r}_1) = \int d^3r_1 h(\vec{r},\vec{r}_1) = 1$.

One should think of normalized Gaussian test functions, for example; the delta function is excluded by condition (ii).

In order to apply Theorem IV, it is only necessary to verify that

(a) $\int J_h(f,f)d^3vd^3r = 0$ (98a)

and

(b) $\nu_h(f) = \int B(\theta)h(\vec{r},\vec{r}_1)f(\vec{r},\vec{v},t)f(\vec{r}_1,\vec{v}_1,t)d\theta d\varepsilon d^3v_1d^3r_1$ (98b)

$L_h(f,f) = \int B(\theta)h(\vec{r},\vec{r}_1)f(\vec{r},\vec{v}',t)f(\vec{r}_1,\vec{v}_1',t)d\theta d\varepsilon d^3v_1d^3r_1$ (98c)

are bounded. We have written here $B(\theta)$ rather than $B(\theta,V)$ since we are considering pseudo-Maxwell molecules.

(a) is easy to show. It is only necessary to mimic the proof given in Chapter I for the case $h \equiv 1$, except that when the change of variable $\vec{v},\vec{v}_1 \leftrightarrow \vec{v}',\vec{v}_1'$ is made, it is also necessary to interchange $\vec{r} \leftrightarrow \vec{r}_1$.

To show (b), compute

$\|\nu(f)\| = \text{ess sup } |\nu(f)(\vec{r},\vec{v},t)| \leqslant hV\sigma\|f\| = hV\sigma\|n\|$. (Recall $V\sigma$ is independent of V). Since h is finite, then $\|\nu\| = \sup\limits_{f} \frac{\|\nu(f)\|}{\|f\|} \leqslant hV\sigma$. (This result makes it clear why the limit $h \to \delta(\vec{r}-\vec{r}_1)$ is difficult to treat.)

In completely analogous fashion, one computes

$\|L\| \leqslant hV\sigma$.

Thus the conditions of Theorem IV are obeyed, and we conclude that the mollified Boltzmann equation (95) has a global, unique, non-negative solution.

13. POVZNER'S GLOBAL EXISTENCE PROOF.

A. Ja. Povzner published an existence proof for a mollified Boltzmann equation in 1962.[15] There is no reference in his paper to Morgenstern's work or, in fact, to any other of the existence proofs already published by 1962 except for Carleman's,[11] so his work should be considered independent of Morgenstern's.

The Povzner paper is more general than Morgenstern's in that the Maxwell molecule assumption is not made (however, a cutoff is assumed, so that the terms $\nu(n)$ and $L(n,n)$ are independently bounded). Also, Povzner allows for the existence of an external force, which is a more or less trivial generalization. (In fact, let us redefine the free propagator T_t to denote motion along the characteristics of the hamiltonian system

$$\frac{dr_i}{dt} = \frac{\partial H}{\partial p_i} \tag{99a}$$

$$\frac{dp_i}{dt} = -\frac{\partial H}{\partial r_i} \tag{99b}$$

with $H = \frac{p^2}{2m} + U(\vec{r})$. The non-vanishing of $U(\vec{r})$ can then be accommodated by using T_t in place of U_t wherever it appears.)

Povzner's proof, then, applies to the derived integral equation

$$n(\vec{r},\vec{v},t) = T_t f_0(\vec{r},\vec{v}) + \int_0^t T_{t-\tau} J(n,n)(\tau)d\tau \tag{100}$$

where f_0 is, as before, the initial datum and the collision integral, $J(n,n)$ involves a mollifier. Boundary conditions, as usual, are assumed to be incorporated into T_t which, by virtue of Lioville's Theorem, is still a unitary group.

$$J(n,n) = \int [n(\vec{r},\vec{v}',t)n(\vec{r}_1,\vec{v}_1',t) - n(\vec{r},\vec{v},t)n(\vec{r}_1,\vec{v}',t)]K(\vec{r}-\vec{r}_1,\vec{V})d^3r d^3v_1. \tag{101}$$

The kernel K ($\mathcal{J}$ in Povzner's notation) mollifies both the spatial and velocity integrations. The usual integral over $d\theta$ is assumed to be implicitly redundant by energy and momentum conservation, and here is omitted. Since Povzner deals only with Eq. (100), he only proves existence of mild solutions, without attempting to prove that they are differentiable.

The kernel K is more or less arbitrary, except that it is assumed to obey certain conditions: $|K(\vec{r},\vec{V})| < M(V+1)$ for some constant M, $K(-\vec{r},-\vec{V}) = K(\vec{r},\vec{V})$, and $\exists\,\rho > 0$ s.t. $K(\vec{r},\vec{V}) = 0$ for $r > \rho$. The first of these conditions, incidentally, guarantees that the total cross section will be finite, so it is equivalent to assuming a "cut-off."

Povzner's work is very technical, containing vastly more computational detail than Morgenstern's, so only the main results will be quoted, along with a sketch of the procedure.

We suppose that $f_0 \geqslant 0$ is normalized to unity[23]

$$\iint d^3v\,d^3r\,f_0(\vec{r},\vec{v}) = 1 \tag{102}$$

and consider Eq. (100) on a very small time interval $[0,\delta]$. Then the equation can be approximated by

$$\hat{n}(\vec{r},\vec{p}) = T_\delta f_0 + \delta T_\delta J(\hat{n},\hat{n}) \tag{103}$$

where $\hat{n}$ represents some average value on the interval $[0,\delta]$. Define an iteration scheme by

$$n_0 = f_0 \tag{104a}$$

$$n_{k+1} = T_\delta f_0 + \delta T_\delta J(n_k,n_k). \tag{104b}$$

Then we state without proof

LEMMA 1. For δ sufficiently small, $n_k \geqslant 0\ \forall k$ (the proof depends crucially on the boundedness of K).

LEMMA 2. If $\int Hf_0 d^3r d^3p \equiv \alpha_2 < \infty$, then $\int Hn_k d^3r d^3p = \alpha_2$.

Here H is the hamiltonian introduced earlier and we have written (and will write hereafter) $\vec{p}$ instead of $\vec{v}$ as is usual in dealing with hamiltonian systems. The proof depends on the fact that $T_h H = H$.

LEMMA 3. The set of functions $n_k(\vec{r},\vec{p})$ is weakly equicontinuous. That is, $\forall h \in L^\infty$, $|\int h(\vec{r},\vec{p})(n_{k+q}(\vec{r},\vec{p}) - n_k(\vec{r},\vec{p}))d^3r d^3p$ can be made small for $q < \varepsilon_k/\delta$, where ε_k depends only on h.

Next, the function $n_\delta(\vec{r},\vec{p},t)$ is constructed as

$$n_\delta(\vec{r},\vec{p},t) = n_k(\vec{r},\vec{p}),\ k\delta \leqslant t \leqslant (k+1)\delta,\ k = 0,1,2,\cdots.$$

Finally,

THEOREM V. Let $f_0 \geqslant 0$ obey Eq. (102) and have finite α_2 (as defined in

Lemma 2). Then ∃ a completely additive measure $\omega(\vec{r},\vec{p},t)$, weakly continuous in t, which satisfies the Boltzmann equation. Furthermore

$$\int d\omega(\vec{r},\vec{p},t) \leqslant 1, \quad \int H(\vec{r},\vec{p})d\omega(\vec{r},\vec{p},t) \leqslant \alpha_2.$$

We note that a measure satisfies the Boltzmann equation (or rather, the weak form (100)) if ∀ measurable h

$$\int h(\vec{r},\vec{p})d\omega(\vec{r},\vec{p},t) = \int h(\vec{r},\vec{p})T_t f_0(\vec{r},\vec{p})d^3rd^3v$$
$$+ \int_0^t d\tau \iint [T_{t-\tau}(h(\vec{r},\vec{p}') - h(\vec{r}_1,\vec{p}_1))K(\vec{r}-\vec{r}_1,\vec{V})d\omega(\vec{r},\vec{p},\tau)d\omega(\vec{r}_1,\vec{p}_1,\tau). \qquad (105)$$

In other words, Eq. (100) is multiplied by $h(\vec{r},\vec{p})$, integrated and $n(\vec{r},\vec{p},t) \times d^3rd^3p$ is replaced by $d\omega(\vec{r},\vec{p},t)$. Povzner comments that such a replacement is not possible for the unmollified equation, and thus expresses doubt that an existence theorem for the unmollified, or otherwise restricted equation, can be proved. (Certainly the events of the past twenty years have not proved Povzner wrong.) Povzner goes on to prove

<u>LEMMA 4</u>. Define $\alpha_k = \int (H(\vec{r},\vec{p}))^{k/2} f_0(\vec{r},\vec{p})d^3rd^3p$. Then if $\alpha_k \leqslant C < \infty$ for $k = 0,1,2,3$, for some constant C, then $\int H(\vec{r},\vec{p})d\omega(\vec{r},\vec{p},t) = \alpha_2$, i.e. total energy is conserved.

Finally, with an additional condition, the measure can be expressed as an L^1 function. so that a less unpleasant solution is obtained. Further, this function is a unique solution to Eq. (100).

<u>THEOREM VI</u>. Suppose for $0 \leqslant t \leqslant \theta$, $\int (H(\vec{r},\vec{p}))^{k/2} d\omega(\vec{r},\vec{p},t) \leqslant C' \leqslant \infty$, $h = 0,1,2,3,4$, for some constant C′ where for $0 \leqslant t \leqslant \theta$ the measure $\omega(\vec{r},\vec{p},t)$ satisfies Eq. (105). Then $\omega(\vec{r},\vec{p},t)$ is the only measure satisfying Eq. (105) and furthermore ∃ a non-negative function $\psi(\vec{r},\vec{p},t)$ s.t. $d\omega(\vec{r},\vec{p},t) = \psi(\vec{r},\vec{p},t)d^3rd^3p$.

"Let us be moral. Let us contemplate existence."

Charles Dickens, Martin Chuzzlewit

V. EXISTENCE THEORY FOR SYSTEMS CLOSE TO EQUILIBRIUM

14. THE BASIC STRATEGY

The fundamental difficulty in attempting to deal with the Boltzmann equation (12) has to do with the fact that the collision term J: $X \to Y$, where X and Y are two distinct Banach spaces. For special interactions, such as the Maxwell molecule model considered so extensively in these lectures, it is possible to choose $X = Y = L^\infty(R^3 \times R^3)$ (or better, as $L^\infty(\Omega \times R^3)$ where the spatial domain $\Omega \not\subseteq R^3$) but, as has already been pointed out in Sec. 7, this space is unnatural from a physical point of view, and this, in some sense, is the reason that only local solutions can be generated. Of course, for the spatially homogeneous problem, as treated in Chapter II, it is possible to choose $X = Y = L^1(R^3)$, and the difficulty evaporates, as we have seen (as it does when a spatial mollifier is introduced, as in Chapter IV). We have already pointed out the difficulty of working in $L^1(\Omega \times R^3)$ for the full spatially-dependent problem, and yet it seems apparent that no other space can lead to a global existence proof for the full spatially dependent Boltzmann equation.

Referring to the "mild" problem, i.e. the integral formulation of the Boltzmann equation

$$n(\vec{r},\vec{v},t) = U_t f_0(\vec{r},\vec{v}) + \int_0^t U_{t-\tau} J(n,n)(\tau)d\tau,$$

if somehow the group U could be used to "kill off" the singularity in J (i.e., if U: $Y \to X$) then we could seek at least mild solutions in $X = L^1$. However, U being a unitary group obviously does not "kill off" anything. However, if we add a further restriction, namely that the initial distribution f_0 is close (in some sense to be defined) to a Maxwell-Boltzmann distribution $M(v)$, then we can stay in one space and global existence proof can be given.

In this perturbation approach, we write[24]

$$n(\vec{r},\vec{v},t) = M(v) + M^{\frac{1}{2}}(v)f(\vec{r},\vec{v},t). \tag{106}$$

Substitution into Eq. (12) leads to

$$\frac{\partial f}{\partial t} + \vec{v}\cdot\vec{\nabla}f = -L(f) + \nu(V)\Gamma(f) \tag{107}$$

where L(f) is the linear collision operator:

$$L(f) = -2M^{-\frac{1}{2}}\, J(M, M^{\frac{1}{2}}f) \tag{108a}$$

and

$$\nu\Gamma(f) = M^{-\frac{1}{2}}J(M^{\frac{1}{2}}f, M^{\frac{1}{2}}f), \tag{108b}$$

where[25]

$$J(f,g) \equiv \tfrac{1}{2}\int d\theta d\varepsilon B(\theta,V)[f'g_1' + f_1'g' - fg_1 - f_1g]d^3v_1; \tag{109}$$

we note $J(f,g) = J(g,f)$.

Explicitly

$$\nu(V)\Gamma(f,g) = \tfrac{1}{2}\int(f'g_1' + f_1'g' - fg_1 - f_1g)M^{\frac{1}{2}}(v)d\Omega \tag{110a}$$

where we have adopted Grad's notation[26]

$$d\Omega = B(\theta,V)d\theta d\varepsilon d^3v_1. \tag{110b}$$

Of course, $\Gamma(f) \equiv \Gamma(f,f)$ while

$$\nu \equiv \int M_1 d\Omega. \tag{110c}$$

The reason for splitting Eq. (110a) into the functions ν times Γ is that although ν is, in general, an unbounded function of V (except for Maxwell molecules) $\Gamma(f,g)$ is a bounded form. ν is related to the total cross section [Eq. (53a)] except that it is Maxwellian averaged.

The linear Boltzmann operator, L(f), can be expressed in quasi-conventional form in the following way. Normalizing $\int M(v)d^3v = 1$,

$$-2M^{-\frac{1}{2}}J(M,M^{\frac{1}{2}}f) = \nu(V)f - \int d^3v_1 K(\vec{v},\vec{v}_1)f(\vec{r},\vec{v}_1,t) \tag{111}$$

and

$$\int d^3v_1 K(\vec{v},\vec{v}_1)f(\vec{r},\vec{v}_1,t) \equiv \int(M'^{-\frac{1}{2}}f_1' + M_1'^{\frac{1}{2}}f' - M^{\frac{1}{2}}f_1)M_1^{\frac{1}{2}}d\Omega. \tag{112}$$

The kernel $K(\vec{v},\vec{v}_1)$ is symmetric, i.e.

$$K(\vec{v},\vec{v}_1) = K(\vec{v}_1,\vec{v}). \tag{113}$$

The non-trivial proof will not be given here.[27]

We have already pointed out that $\nu(V)$ is unbounded, except for Maxwell molecules. The term involving the kernel $K(\vec{v},\vec{v}')$ is, on the other hand, very well behaved. In fact, in the appropriate Banach space it is the kernel of a compact linear transformation, K.

The idea now will be to rewrite the Boltzmann equation in the form

$$\frac{\partial f}{\partial t} + \vec{v}\cdot\vec{\nabla} f + Lf = \nu\Gamma(f),\ f(\vec{r},\vec{v},0) = f_0, \tag{114}$$

and to use $\vec{v}\cdot\vec{\nabla} + L$ as the generator of a semigroup, call it G_t. Then the mild form of Eq. (114) can be written

$$f(\vec{r},\vec{v},t) = G_t f_0(\vec{r},\vec{v}) + \int_0^t G_{t-\tau}\nu\Gamma(f(\tau))d\tau \tag{115}$$

and try to show that G kills off the singularity in $\nu\Gamma$ (since $\nu\Gamma$ has inherited the obnoxious properties of J described earlier in this section). We shall see, in fact, that, at least for sufficiently small deviations from the equilibrium (i.e. for sufficiently small f_0) this is indeed the case.

Incidentally,[24] the operator L is non-negative, but it is not positive since clearly $\lambda = 0$ is a five-fold degenerate eigenvalue with eigenvectors $M^{\frac{1}{2}}$, $\vec{v}M^{\frac{1}{2}}$ and $v^2M^{\frac{1}{2}}$ corresponding to conservation of particles, momentum and energy. Furthermore, due to the compactness of K, the essential spectrum consists of the support of $\nu(V)$. Thus $\lambda = 0$ is an isolated eigenvalue.

The intermolecular force law considered here is assumed to be "hard" in the sense of Grad. This means that it is sufficiently strong for small distances of separation. In particular, for inverse power force laws, of the form of Eq. (55), we need $s \geqslant 5$. In addition, the force must be repulsive and of finite total cross section (this means the usual cutoff has been introduced). Even though $\nu(V)$ is unbounded, for hard potentials it obeys[28]

$$0 \leqslant \nu_0 \leqslant \nu(V) \leqslant \nu_1(1+V^2)^{\frac{1}{2}} \tag{116}$$

for some constants ν_0 and ν_1.

The results to be described in this chapter are, by and large, based on the pioneering work of H. Grad [24,28,29] who first introduced the linearization (106) and derived the properties of the various linear and non-linear operators $\nu(V)$, K and Γ. In addition, i.a., he proved the following crucial lemma.

<u>LEMMA (GRAD'S INEQUALITY).</u>[26] Assume J(n,n) describes a gas interacting through a cutoff, hard potential. For $\Omega \subset R^3$, define the set

$B_\mu = \{f\colon \Omega \times R^3 \to C \mid f \text{ continuous}, \sup_{\vec{r},\vec{v}} (1+v^2)^{\mu/2}|f(\vec{r},\vec{v})| < \infty\}$, a Banach space with norm $\|\cdot\|_\mu$. Then

$$\|\nu\Gamma(f,g)\|_{\mu-1} \leq C_\mu \|f\|_\mu \|g\|_\mu,$$

where C is a constant.

PROOF. The proof involves three pages of computations in the reference cited, and will not be reproduced in detail. The basic idea is to decompose $\nu\Gamma$, Eq. (110a), into two terms:

$$|\nu\Gamma| \leq \nu\Gamma_1 + \nu\Gamma_2$$

with

$$\nu\Gamma_1 = \int \tfrac{1}{2}|fg_1 + f_1 g| M_1^{\frac{1}{2}} d\Omega$$

and

$$\nu\Gamma_2 = \int \tfrac{1}{2}|f'g_1' + f_1'g'| M_1^{\frac{1}{2}} d\Omega.$$

To estimate Γ_1, one uses the bound

$$\lambda = \int M_1^{\frac{1}{2}} d\Omega < \alpha_0 \nu$$

for some constant α_0. In proving this inequality, the "hard" character of the potential is utilized. This leads directly to the estimate

$$\|\Gamma_1\|_\mu \leq \alpha_0 \|f\|_\mu \|g\|_\mu. \tag{117a}$$

The estimate of Γ_2 is obtained by estimating the velocity integrations involved in $d\Omega$, leading to

$$\|\Gamma_2\|_\mu \leq \gamma_0 \|f\|_\mu \|g\|_\mu \tag{117b}$$

for some constant γ_0. (α_0 is independent of ν, but γ_0 is not.) Recalling that $0 < \nu \leq \nu_1(1+V^2)^{\frac{1}{2}}$, the result follows.

Also, we have commented on the compactness of the linear operator $K = L + \nu$. In particular the following result is useful:

$$\|Kf\|_\mu \leq D_\mu \|f\|_{\mu-1}. \tag{118}$$

We are now equipped to discuss existence theory for systems close to equilibrium.

15. EXISTENCE THEORY FOR SYSTEMS CLOSE TO EQUILIBRIUM. PROOF 1.

The first proof to be presented here was given by Shizuta and Asano,[30] and involved a bounded spatial domain with specular reflection at the walls. In particular, let $\Omega \subset R^3$ be a bounded convex domain, and assume that $\partial\Omega \in C^3$. The function space of choice is $S_\mu \subset B_\mu$, such that $\forall f \in S_\mu$

(i) f is continuous on $\Omega \times R^3$

(ii) for $\vec{r} = \vec{r}_s \in \partial\Omega$, $f(\vec{r}_s,\vec{v}) = f(\vec{r}_s,\vec{v} - 2\vec{n}(\vec{v}\cdot\vec{n}))$, where n is the normal to $\partial\Omega$ at $\vec{r}_s$. (This condition merely states that specular reflection boundary conditions are being considered.

(iii) $\sup_{\vec{r}} (1+v^2)^{\mu/2}|f(\vec{r},\vec{v})| \to 0$ as $v\to\infty$.

Define the operator $Bf = \vec{v}\cdot\vec{\nabla}f + Lf$. We recall that in Sec. 14 we denoted the semigroup generated by B by G_t. We have

PROPOSITION. The semigroup G_t is contractive in S_α, $\alpha \geqslant 0$. Further, the imaginary axis belongs to the resolvent set of B, except for $\lambda = 0$, which is an isolated eigenvalue of B, with eigenvectors $M^{1/2}$, $M^{1/2}v$, $M^{1/2}v^2$.

PROOF. We have already pointed out that the free propagator U_t, which is generated by $\vec{v}\cdot\vec{\nabla} + \nu$, is a unitary group and, hence, a bounded semigroup. But G_t is generated by $\vec{v}\cdot\vec{\nabla} + \nu + K$, where K is a compact perturbation, so by a well-known theorem,[31] G_t is a contractive semigroup. The fact that $\lambda = 0$ is a five-fold degenerate, isolated eigenvalue, was pointed out in Sec. 14.

Next

LEMMA 1. Let P denote the projection onto

$A = Sp(M^{1/2}, M^{1/2}\vec{v}, M^{1/2}v^2)$.

Then $\forall\gamma > 0$, $\exists\, C > 0$ depending only on α and γ s.t.

$\|G_t(1-P)\| \leqslant Ce^{-\gamma t}$, $t \geqslant 0$.

Here $\|\cdot\|$ represents the operator norm, $S_\alpha \to S_\alpha$.

PROOF. G_t is a strict contraction on $(1-P)S_\alpha$.

Before proceeding with the existence proof, let us consider where we stand. Suppose we begin with initial datum $f_0 \in B_\mu$. We could divide the interval $[0,T]$ into small time steps Δt, prove local existence for $0 \leqslant t \leqslant \Delta t$, use $f(\vec{r},\vec{v},\Delta t)$ as the datum for the step $\Delta t \leqslant t \leqslant 2\Delta t$, and proceed interval by interval. Of course, this requires some control on the growth of f in the interval Δt which was available when the L^1 norm was used--Chapters II and IV--but more difficult in the sup norm required for the unmollified spatial problem considered here. We do have on useful result, namely for $f \varepsilon B_\mu$, $\nu\Gamma(f)\varepsilon B_{\mu-1}$ (Grad's inequality). This, along with Lemma 2, below, turns out to provide sufficient control on the growth of f to permit an existence proof, at least for sufficiently small initial datum (which is to say, for systems sufficiently near equilibrium). Using Grad's inequality and Lemma 1 with $\delta < \gamma$ yields

LEMMA 2. Define

$X_{\mu,\delta} = \{f: [0,\infty] \to B_\mu \mid f$ is continuous in the B_μ strong topology, $\|f\|_{\mu,\delta} \equiv \sup_{t \geq 0} e^{\delta t}\|f(t)\|_\mu < \infty\}$; $N_\mu = \{f \varepsilon B_\mu \mid Pf = 0\}$.

Then $\exists\, \delta > 0$ s.t. $f \varepsilon X_{\mu,\delta} \cap N_\mu$, $\int_0^t G_{t-\tau}\nu\Gamma(f(\tau))d\tau \varepsilon X_{\mu,\delta}$.

The idea is to analyze $G = \int_0^t G_{t-\tau}\nu(\cdot)d\tau$ by treating G_t as the sum of $\vec{v}\cdot\vec{\nabla} + \nu$ and K and use the Duhamel formula to obtain $G: B_\mu \to B_\mu$.

Now we can state

<u>THEOREM VII.</u> Let $f_0(\vec{r},\vec{v})\varepsilon \mathcal{S}_\mu \cap N_\mu$, $\|f_0\|_\mu = E$. Then for E sufficiently small $\exists!$ solution to Eq. (115).

<u>PROOF.</u> Define an iterative scheme for Eq. (115) by

$$f^{(-1)}(t) = 0$$

$$f^{(n)}(t) = G_t f_0 + \int_0^t G_{t-\tau}\nu\Gamma(f^{(n-1)}(\tau))d\tau.$$

Then by Lemma 2, $f^{(n)}\varepsilon X_{\mu,\delta}$ for some $\delta > 0$ and $\forall t > 0$.

$$\|f^{(n)}\|_{\mu,\delta} \leqslant CE + C_1\|f^{(n-1)}\|^2_{\mu,\delta}$$

where C and C_1 are constants. It follows easily that this scheme converges in $X_{\mu,\delta}$ for $E < \frac{1}{4CC_1}$. Uniqueness is evident.

<u>REMARK.</u> Since $f(t)\varepsilon X_{\mu,\delta}$, $f(t)\to 0$ exponentially as $t\to\infty$, i.e. the solution approaches the equilibrium Maxwellian distribution [cf. Eq. (106)]. Note also that not only is the norm of the initial datum assumed small, but it is assumed that $f_0 \varepsilon N_\mu$, i.e. all the conserved moments of f_0 vanish. Thus, the fluid characteristics of this model are all carried by the unperturbed solution, i.e. the Maxwellian M(v). For this reason, this result is somewhat deficient.

16. <u>OTHER NEAR-EQUILIBRIUM EXISTENCE PROOFS.</u>

A number of other authors have considered the existence of solutions for systems close to equilibrium. In this section, we shall attempt to sketch some of the cogent results, without going into too much detail.

Nishida and Imai[32] avoid the somewhat objectionable assumption $f_0 \in N_\mu$ Ref. 30. However, they consider $\Omega = R^3$. Denote by H^p, $p \geqslant 0$ the Sobolev space of order p defined by

$$H^p = \{f: R^3 \to R \mid \frac{d^n}{dx_i^n} f \in L^2(R^3),\ 0 \leqslant n \leqslant p,\ i = 1,2,3\}.$$

H^p is a Hilbert space with norm

$$\|f\|_{H^p} = |(1+k)^p \hat{f}(k)|_{L^2}$$

where $\hat{f}$ is the Fourier transform of f.

Next, define the Hilbert space

$$H_p = L^2(\vec{v}, H^p(x)),\ \text{i.e.}\ H_p = \{f: R \to H^p \mid \|f\|^2 = \int \|f(\vec{v},\cdot)\|^2_{H^p} d^3v\}.$$

Furthermore, define $B_{m,p} \subseteq H_p$, $m,p \geqslant 0$ to consist of strongly continuous functions s.t. $(1+|v|^m)\|f(\vec{v},\cdot)\|_{H^p} \to 0$, $|v| \to \infty$ and $\|f\|_{m,p} = \sup_{\vec{v}} (1+|v|^m)\|f(\vec{v},\cdot)\|_{H^p}$.

For $m > \frac{5}{2}$, $p > \frac{3}{2}$, a Grad-type inequality can be proved for $B_{m,p}$:

$$\|\nu\Gamma(f,g)\|_{H_p} \leqslant C\|\Gamma(f,g)\|_{m,p} \leqslant \hat{C}\|f\|_{m,p}\|g\|_{m,p}$$

where C and $\hat{C}$ are constants. Thus, we proceed as in Sec. 15. Let $C(R^+,B_{m,p})$, denote the space of continuous functions $f: R^+ \to B_{m,p}$ which decay to zero in $B_{m,p}$ as $t \to \infty$. Then $\|G_t f\|_{m,p} \leqslant C_1\|f\|_{m,p}$ and $\forall f \in B_{m,p}$, $G_t f \in C(R^+,B_{m,p})$. Define

$$h(t) = \int_0^t G_{t-\tau}\Gamma(f(\tau))d\tau.$$

Then $h(t) \in C(R^+,B_{m,p})$ if $f \in C(R^+,B_{m,p})$, $m \geqslant \frac{3}{2}$, $p \geqslant 2$. This is the "killing off" property of G_t. There follows

<u>THEOREM VIII</u>. If $f_0 \in B_{m,p}$, $m \geqslant \frac{3}{2}$, $p \geqslant 2$, and $\|f_0\|_{m,p}$ is sufficiently small, $\exists!$ global mild solution to the Boltzmann equation s.t. $f(t) \to 0$ in $B_{m,p}$ as $t \to \infty$.

Additional estimates on the decay of f(t) for large t can be obtained if further restrictions are placed on the initial datum. The reader is referred to Ref. 32 for details.

A slight improvement on the treatment of Ref. 32 was made by Guiraud,[33] who considered general boundary conditions on $\partial\Omega$ of the type $f_{out} = K_0 f_{in}$, where K_0 is some linear operator. The proof is more complicated than that sketched above, but the general idea is the same.

Finally, Shizuta[34] has considered classical solutions, i.e. regular (not distributional) solutions of the differential equation (rather than mild solutions). The basis of this work (which allows one to deduce the existence of classical solutions for the situations considered above) is the following

LEMMA. Consider the differential equation $x'(t) = Ux + F(x)$, $x(0) = x_0$. Assume $X \subseteq Y$ a continuous dense imbedding, let U generate a strongly continuous semigroup T_t in X and in Y and $x \to F(x)$ is continuous from X to Y. Finally, let $X \subseteq D(U) \subseteq Y$ be continuous imbeddings where $D(U)$ on Y is equipped with the graph norm and $U: X \to Y$ is continuous from X to Y. Then if $x \in C(R^+,X)$ satisfies the mild equation

$$x(t) = T_t x_0 + \int_0^t T_{t-\tau} F(x(\tau))d\tau, \; x(0) = x_0 \in X,$$

then $x \in C^1(R^+,Y)$ and $x(t) \in D(U)$ for $t \geq 0$. Moreover, x is a strict solution of the differential equation in Y.

Note: A strict solution is a classical solution if the derivatives are taken in the classical sense.

Finally, we mention some earlier work of Ukai.[35] This work was of a seminal nature, since most of the techniques we have been describing here were anticipated by Ukai. However his results were somewhat weaker inasmuch as his method could not accommodate boundaries, and had various other inadequacies, particularly with reference to smoothness of the solutions.

"Cogito, ergo sum."

R. Descartes, Le discours de la méthode

VI. LATTICE SYSTEMS

17. GENERAL CONSIDERATIONS.

We have already discussed in detail (Sec. 11) the motivation for introducing a mollifier into the collision term of the Boltzmann equation. We also mentioned the possibility of an alternate procedure (also inspired by the example of quantum field theory[20]), namely setting the Boltzmann equation on a spatial lattice. One can prove existence for non-zero lattice spacing,[36,37] and some progress has been made in studying the existence of a limit as the lattice spacing tends to zero.[38]

We follow here basically the method described in Refs. 36 and 38. The Boltzmann equation is treated in the real Banach space

$$B^n = L^1(\Gamma_n \times R^3)$$

with norm

$$\|f\|_{B^n} = 2^{-3n}\sum_{i\varepsilon\Gamma_n}\int_{R^3}|f_i|d^3v.$$

Here, $\Gamma_n = \{1\cdot 2^{-n}, 2\cdot 2^{-n}, \ldots, 2^{3n}\cdot 2^{-3n}\}$ is a discrete set with cardinality 2^{3n} representing a lattice approximation to the continuum $\Gamma = [0,1]^3$, with lattice spacing 2^{-n}. On the boundary points of Γ_n, periodic boundary conditions are imposed, corresponding in a continuous model to specular reflection.[39]

The gradient term of the Boltzmann equation is replaced by a finite-difference operator A^n. Thus

$$2^{-n}(A_x^n f)_{(j,k,l)} = \begin{cases} v_x(f_{(j,k,l)} - f_{(j-2^{-n},k,l)}), & v_x > 0 \\ v_x(f_{(j+2^{-n},k,l)} - f_{(j,k,l)}), & v_x < 0 \end{cases}$$

with similar expressions for A_y^n and A_z^n. Observe that for $v_x > 0$ a forward difference is used in the definition of A_x^n, while for $v_x < 0$ a backward difference is used. This is a more or less standard technique in numerical analysis, and is required in order to preserve positivity.

If we think of A^n as a $2^{3n} \times 2^{3n}$ matrix and the dependent variable $\{n_i(\vec{v})\}$ as a 2^{3n}-column vector, then the 2^{3n} lattice sites can be enumerated in such a way that A^n can be constructed by tensor products.

$$A^n = (v_x + v_y + v_z) I^n \otimes I^n \otimes I^n - v_x (E \otimes I^n \otimes I^n) - v_y (I^n \otimes E^n \otimes I^n)$$
$$- v_z (I^n \otimes I^n \otimes E^n), \quad v_x, v_y, v_z, \geqslant 0. \tag{119}$$

where I^n is the identity matrix and E^n is defined by

$$E^n_{ij} = \begin{cases} \delta_{n,j}, & i = 1 \\ \delta_{i,j+1}, & i > 1 \end{cases} \tag{120}$$

If $v_x < 0$, then the corresponding A^n is obtained from (119) by the replacement $E^n \to E^{n*}$ in the second term, etc. (For simplicity of notation, we hereby drop the superscript n, reintroducing it as necessary in Sec. 17.) Physically, this model corresponds to a Markov process in which a particle at point (i,j,k) "jumps" to the point $(i+2^{-n},j,k)$ with a probability proportional to v_x/v, $v_x > 0$ (or to the point $(i-2^{-n},j,k)$ for $v_x < 0$), and similarly for the motions in the y and z directions.[37] We denote by D(A) the domain of the unbounded operator A.

The differential form of the Boltzmann equation which we consider is

$$\frac{\partial n_i(\vec{v},t)}{\partial t} + (An)_i = J(n_i,n_i), \quad n \in B, \quad n_i(\vec{v},0) = f_{i0}(\vec{v}), \tag{121}$$

and we shall assume the operator J to correspond to a bounded, cut-off force law (e.g. pseudo-Maxwell molecules) so that writing as usual

$$J(\phi,\phi) = L(\phi,\phi) - \nu(\phi)\phi \tag{122}$$

both L and ν are bounded, continuous transformations.

We recall that we have introduced (Sec. 8) the free propagator U_t (generated by $\vec{v}\cdot\vec{\nabla}$) and the "destructive" propagator D_t (generated by $\vec{v}\cdot\vec{\nabla} + \nu$). In the present case because of the matrix nature of the problem being considered, D_t must be replaced by an evolution group, call it $T(t_1,t_2)$ (the physicist's "time-ordered exponential"). Observe that T depends on n; we may write $T(n;t_1,t_2)$.[38a]

We shall consider mild solutions generated by U_t and $T(n;t_1,t_2)$ (we continue to use the notation U_t although the definition of U_t differs from that given in Eq. (80a).

$$n(t) = U_t f_0 + \int_0^t U_{t-\tau} J(n,n)(\tau)d\tau \tag{123a}$$

and

$$n(t) = T(n;t,0)f_0 + \int_0^t T(n;t,\tau)L(n,n)(\tau)d\tau \tag{123b}$$

and use the following technique. First, local existence in B^n will be proved for Eqs. (123a) and (123b) by setting up an iteration scheme and proving that the sequence of iterates is Cauchy (fixed point arguments can be used instead[37]). Next, it will be shown that any solution of (123a) is also a solution of (123b). Because all terms of (123b) are positive for $f_0 > 0$, the usual conservation laws (Sec. 3) allow us to prove that $\|n(t)\|_{B^n}$ is a constant. Thus, the solution at the end of the "sufficiently small" time interval can be used to generate the solution for the next time interval, and so on ad infinitum. Finally, an argument based on holomorphic semigroups is used to prove that the solution is differentiable and hence is actually a solution of Eq. (121).

The details of the proof are sketched in the next section. In Sec. (19) the existence of a limiting solution, as $n \to \infty$ (lattice spacing tends to zero) is discussed.

18. EXISTENCE AND UNIQUENESS PROOF--LATTICE APPROXIMATIONS.

We define $B^+ \subset B$ as the cone of positive functions and $G(B^+)$ the cone of measurable functions $\psi(\cdot)$: $R^+ \to B^+$. A solution of Eq. (121) (or (123)) means a strongly differentiable map n: $R^+ \to B$ such that $f_0(\vec{v}) \in B^+ \cap D(A)$ (the restriction $f_0 \in D(A)$ can be removed[37]).

Crucial to the existence proofs are the following properties of U_t and $T(n;t_1,t_2)$:

LEMMA 1.

(a) U_t and $T(n;t_1,t_2)$ are invariant on B^+ for t, t_1-t_2 positive and $n \in G(B^+)$

(b) U_t is a contraction semigroup and continues analytically to a bounded holomorphic semigroup $U(\xi)$.[39]

(c) $T(n;t_1,t_2)$ is a contraction mapping on B for $t_1-t_2 > 0$ and $n \in G(B^+)$.

PROOF. Part (a) is proved by working on the dense subspace M_0 of functions in B^+ with compact support, and expressing $U_t = e^{-At}$ as a power series on M_0. Similarly, $T(n;t_1,t_2)$ has an explicit representation as a product of exponentials over partitions of $[t_1,t_2]$.[40] Applying the Lie product formula[41] to the product of exponentials yields the desired result. (b) follows from the simple fact that a B^+-invariant contraction is also a contraction on the whole space. The analyticity is proved by observing that $U(\xi) = U_x(\xi) \otimes U_y(\xi) \otimes U_z(\xi)$ and then obtaining an explicit expression for $U_x(\xi) = e^{\xi(I-E)}$ on H_0. From this expression, the analytic properties of $U_x(\xi)$ (and hence $U(\xi)$) can be read off, leading to the conclusion.

To prove (c), it is only necessary to show that T is the product of positive exponentials. This follows from the Ginzburg representation.[40]

COROLLARY 1. $\forall f \in B, \sum_{i=1}^{2^{3n}} (U_t n)_i = \sum_{i=1}^{2^{3n}} n_i$.

This is the statement that the lattice approximation preserves translational invariance. In the continuous case the analogous statement is

$\int d^3r n(\vec{r},\vec{v},t) = \int d^3r n(\vec{r}-\vec{v}t,\vec{v},0)$.

COROLLARY 2. $\rho(t) \equiv \int \sum_i d^3v n_i(\vec{r},\vec{v},t) = \rho$, where ρ is a constant independent of time.

PROOF. Sum Eq. (123b) over i and integrate over d^3v using Corollary 1 and the fact that 1 is a collision invariant (cf. the proposition on p. 14).

REMARKS. If $n \geqslant 0$ Corollary 2 implies that $\|n\|(t) = \|f_0\|$. The essence of our proof will involve showing the positivity of n(t), and then to use the conservation of norm.

We now solve Eq. (123b) by iteration, defining

$$n^{(0)}(\vec{v},t) = f_0(\vec{v}) \tag{124a}$$

and

$$n^{(k)}(\vec{v},t) = U_t f_0(\vec{v}) + \int_0^t U_{t-\tau} J^{(k-1)}(\tau)d\tau \tag{124b}$$

with

$J^{(k)} \equiv J(n^{(k)},n^{(k)})$. Then

LEMMA 3. For t sufficiently small, $\|n^{(k)}(t)\| \leq M$, independent of t and n.

The proof is immediate, since from Eq. (124b) $\|n^{(k)}\| \leq \|f_0\| + t\|J\|\|n^{(k-1)}\|^2$, so the result follows for $t < (4\|J\|\|f_0\|)^{-1}$.

Similarly, the sequence $\{n^{(k)}\}$ is seen to be Cauchy from the estimate

$$\|n^{(k+1)}-n^{(k)}\| \leq t\|J\|M\|n^{(k)}-n^{(k-1)}\|$$

so that

LEMMA 4. The iterative scheme (124) converges to a solution n(t) of Eq. (123b) for $t < \max\{(4\|J\|\|f_0\|)^{-1}, (2\|J\|M)^{-1}\}$, and n(t) is a continuous function of f_0.

Next set up an iterative scheme to Eq. (123a) by

$$n^{(0)}(\vec{v},t) = f_0(\vec{v}) \tag{125a}$$

$$n^{(p)}(\vec{v},t) = T(n^{(p)};t,0)f_0(\vec{v}) + \int_0^t T(n^{(p)};t,\tau)L^{(p)}(\tau)d\tau, \tag{125b}$$

with $L^{(p)}(\tau) \equiv L(n^{(p)}(\vec{v},\tau),n^{(p)}(\vec{v},\tau))$.

Then

LEMMA 5. Define $n^{(p)}(\vec{v},t)$ by the iteration scheme (125). Then for t sufficiently small, $n^{(p)}$ converges in B to $n(\vec{v},t)$, a solution of Eq. (123a). Further, for $f_0 \in B^+$, $n(\vec{v},t) \in G(B^+)$.

The proof is slightly technical, but is again based on proving the sequence of iterates is Cauchy in B using Lemma (1c). Actually, a "fractional" iteration scheme is used, where $n^{(p+\frac{1}{2})}$ is obtained by improving $L^{(p-1)}$ to $L^{(p)}$ and $n^{(p+1)}$ is then obtained by improving $T(n^{(p-1)})$ to $T(n^{(p)})$. Positivity follows from Lemma (1a). We now have the crucial result:

LEMMA 6. Let $n_1(\vec{v},t)$, $n_2(\vec{v},t)$ be solutions of Eqs. (123a) and (123b) respectively satisfying $n_1(\vec{v},0) = n_2(\vec{v},0) = f_0(\vec{v})$. Then $n_1 = n_2$.

PROOF. We omit the somewhat technical details. As usual, the idea is to show that the difference n_1-n_2 is identically zero. Gronwall's Lemma plays an important part in the demonstration.

Immediate is

COROLLARY. The solution to Eqs. (123a), (123b) is unique.

The following theorem of Kato[39] is now used to prove the mild solutions so far obtained are actually strong solutions.

THEOREM (KATO). Let T be the generator of a holomorphic semigroup U_t on a Banach space X and f: $R^+ \to X$ a Hölder continuous function. Then the equation

$$\frac{d\phi(t)}{dt} = -T\phi(t) + f(t), \quad t > 0$$

subject to $\phi(0) = f_0 \in X$, has the solution

$$\phi(t) = U_t f_0 + \int_0^t U_{t-s} f(s)ds$$

and $\phi(t)$ is continuously differentiable for $t > 0$.

Referring to Eq. (123a), it is sufficient to show that J(n,n) is Hölder continuous since we have already shown (Lemma 1(b)) that U_t continues to a holomorphic semigroup. By Hölder continuous, we mean $\|J(t) - J(s)\| \leq M|t-s|^{\alpha}$ for some constants M and $\alpha \leq 1$. It is not too hard to see that the Hölder continuity of J follows from that of n(t). Thus

LEMMA 7. For $f_0 \in B^+ \cap D(A)$, $\{n^{(k)}(t)\}$ as specified by the iterative scheme (124) are differentiable on some interval $[0,T_0]$, and the derivatives $n^{(k)\prime}(t)$ are uniformly bounded (in k and t).

PROOF. Computation.

COROLLARY. The $\{n^{(k)}(t)\}$ are uniformly Lipschitz (and hence Hölder) continuous on $[0,T_0]$.

Now

THEOREM IX. For $f_0 \in B^+ \cap D(A)$, the solution of Eq. (123b) is differentiable, and hence a solution of (121), $t < T_0$.

PROOF. All the conditions of Kato's theorem are satisfied.

Now

LEMMA 8. For $f_0 \in B^+ \cap D(A)$, $\|n(t)\| = \|f_0\|$ for t sufficiently small that n(t) obeys both (123a) and (123b).

PROOF. See the remarks following Corollary 2 to Lemma 1 and recall (Lemma 5) that n(t) is positive.

Now the global property of the solution $n(t)$ follows immediately, since the solution can be obtained in sufficiently small time steps, and this procedure, by virtue of Lemma 8, can be carried out ad infinitum (we observe the "sufficiently small" times we are dealing with depend only on $\|f_0\|$, a constant M independent of t and $\|J\|$. Thus

THEOREM X. Suppose $f_0 \varepsilon B^+ \cap D(A)$. Then $\exists!$ positive solutions $n(t)$ of the integral equations (123a) and (123b) $t \geq 0$ and $n(t)$ is a continuously differentiable solution of Eq. (122) $t > 0$, $n(t) \varepsilon G(B^+)$, $t \geq 0$ and $n(t)$ depends continuously on the initial datum f_0.

Spohn[37] has proved a similar result by a different method (actually he is able to dispense with the condition $f_0 \varepsilon D(A)$, so his result is somewhat more general).

Spohn's technique has some similarities to that of Ref. 17, in that he considers a linear equation of the type

$$\frac{\partial n_i}{\partial t} + (An)_i = J(n_i(\vec{r},\vec{v},t),\hat{n}_i(\vec{r},\vec{v})). \tag{126}$$

This equation is solved for a time interval $0 \leq t \leq t$, with initial datum $n(\vec{r},\vec{v},0) = f_0(\vec{r},\vec{v})$, and $\hat{n}(\vec{r},\vec{v}) = f_0(\vec{r},\vec{v})$. For the second time interval $t_1 \leq t \leq t_2$, Eq. (126) is solved with $\hat{n}(\vec{r},\vec{v}) = n(\vec{r},\vec{v},t_1)$ with initial datum $n(\vec{r},\vec{v},t_1)$, etc. For $t_j - t_{j-1}$ small, the solution of Eq. (126) approaches the solution of the non-linear Boltzmann equation (122). In fact Spohn expresses the solution of Eq. (122) as a fixed point of a certain mapping Z, and proves that Z has a unique fixed point. Defining the operator

$$B^{(\hat{n})}(n) = J(n,\hat{n})$$

then $Z\colon n(t) \mapsto T(\hat{n};t,0)$, where T is the same evolution group considered earlier in this section. A theorem of Voigt[42] permits relaxing the restriction $f_0 \varepsilon D(A)$.

The question of the existence of a limiting solution as the lattice spacing tends to zero is only partially answered at the present time. The efforts which have been carried out so far are described in the next section.

19. LATTICE LIMIT AND NON-MAXWELLIAN GASES.[38]

In the present section we prove the existence of a weak limit to the solutions of the lattice Boltzmann equation described in the previous section as the lattice spacing tends to zero ($n\to\infty$). In addition, the collision model is generalized to some extent. The basic idea is to restrict the initial datum to finite energy and entropy, and to show that the lattice approximation conserves energy and satisfies an H-theorem with the (increasing) entropy and energy uniformly bounded in n. This permits the application of weak compactness arguments similar to those introduced in Refs. 13 and 14(a),(b). We recall (Sec. 3) that

$\int J(f,f)d^3v = \int v^2 J(f,f)d^3v = 0.$ In addition

PROPOSITION. $\int J(f,f)\ln f d^3v \leq 0$. (We now use f to represent the dependent variable of the Boltzmann equation to avoid confusion with the index n representing the lattice spacing.)

This inequality implies the H-theorem proved heuristically in Sec. 3. A rigorous proof can be formulated, subject to some technical restrictions--cf. Ref. 41. Some of the technical details are sketched in the Appendix.

Define now the Banach space $B^n_k = \{f \in B^n | (1+v^2)^{k/2} f \in B^n\}$ with norm

$$\|f\|_k = \|(1+v^2)^{k/2} f\|_{B^n}. \tag{127}$$

Then

LEMMA 1. (a) Let $f_0 \in B^{n+}$ and let $f(t)$ be a solution to Eq. (123a) with $f(0) = f_0$. Then $f(t) \in B^{n+}$ and $\|f(t)\|_{B^n} = \|f_0\|_{B^n}$.

(b) Let $f_0 \in B^{n+}_2$, and $f(t)$ as above. Then $\|f(t)\|_2 = \|f_0\|_2$.

Part (a) was proved in Sec. 18 (Lemma 8) under the assumption $f_0 \in B^{n+} \cap D(A)$. The restriction $f_0 \in D(A)$, as we have noted, was removed by Spohn.[37] To prove part (b), mimic the existence proof of Sec. 18 (Ref. 36) for the space B^n_2 to conclude that for $f_0 \in B^{n+}_2$ $\exists!$ solution $f(t) \in B^{n+}_2$. Then multiplication of Eq. (123a) by v^2 and integration over d^3v yields the desired result.

LEMMA 2. Let $f_0 \in B^{n+}$, $f_0 \ln f_0 \in B^n$, and define

$$H(f_0) = \sum_{i\varepsilon\Gamma_n} \int d^3v f_{0i}(\vec{v}) \ln f_{0i}(\vec{v}).$$

Then $H(U_t^n f_0(\vec{v}))$ is a non-increasing function of t. Here U_t^n is the semigroup generated by A^n(i.e. n is an index, not an exponent).

PROOF. Define the matrix elements $U_{t,ij}^n(\vec{v})$ by $(U_t^n f)_i(\vec{v}) = \sum_{j\varepsilon\Gamma_n} U_{t,ij}^n(\vec{v}) f_j(\vec{v})$. Then the $U_{t,ij}^n$ are all positive. (From the positivity of U_t^n, Lemma 1(a), Sec. 18, positivity follows for sufficiently small t; the semigroup property $U_{t_1}U_{t_2} = U_{t_1+t_2}$ allows extension to all t.) By Lemma 1(c) $\sum_{i\varepsilon\Gamma_n} U_{t,ij}^n(\vec{v}) = 1$. Considering the lattice points as state space, and fixing $\vec{v}$, the matrix $U_t^n(\vec{v})$ is seen to be the transition matrix for a discrete Markov system. Since any space-independent distribution $\psi(t) \varepsilon B^n$ is a fixed point of $U_t^n(\vec{v})$, standard arguments[43] prove that $H(U_t^n(\vec{v})f_0)$ is non-increasing. In particular, writing $\alpha(x) = x\ln x$ and utilizing the convexity of α, we obtain

$$\sum_{i\varepsilon\Gamma_n} \int d^3v \alpha((U_t^n(\vec{v})f_0)_i) \leqslant \sum_{i,j\varepsilon\Gamma_n} \int d^3v U_t^n(\vec{v})_{ij} \alpha(f_{0j}) = \sum_{j\varepsilon\Gamma_n} \int d^3v \alpha(f_{0j}).$$

LEMMA 3. Let $f_0 \varepsilon B_2^{n+}$, $f_0 \ln f_0 \varepsilon B^n$. Let f(t) be a solution to (123b) with $f(0) = f_0$. Then $f(t)\ln f(t) \varepsilon B^n$, $t \geqslant 0$ and $H(f(t))$ is a non-increasing function of t.

PROOF. This follows from a theorem of Voigt.[44] (see Appendix).

Now, we have the tools available to prove weak compactness. Define the Banach spaces

$$B = L^1(\Gamma \times R^3)$$

and

$$B_T = L^1([0,T] \times \Gamma \times R^3)$$

where, we recall, $\Gamma = [0,1]^3$. The spaces B_k and $B_{T,k}$ are defined analogously to the space B_k^n (cf. Eq. (127)). Introduce the projections P^n: $B \rightarrow B^n$ by

$$(P^n(f))_i(\vec{v}) = 2^{3n} \int_{\Delta_i} f(\vec{r},\vec{v}) d^3r \qquad (128a)$$

where Δ_i is a cubical plaquette of side 2^{-n} associated with the ith lattice point. Define also the injections I^n: $B^n \rightarrow B$ by

$$(I^n f)(\vec{r},\vec{v}) = f_i(\vec{r}), \ \vec{r} \in \Delta_i. \qquad (128b)$$

Then

LEMMA 4. Let $f_0 \in B_2^+$ s.t. $f_0 \ln f_0 \in B$. Let $f^n(t)$ be the solution of Eq. (123a) with $f^n(0) = P^n f_0$. Then $H(f^n(t)) \leqslant H(f_0)$ and

$$\|f^n(t)\|_{B_2^n} = \|f_0\|_{B_2}, \ t \geqslant 0.$$

PROOF. Follows from Lemma 3 and Jensen's inequality.[45]

We can now state

THEOREM XI. Let $f^n(t)$ be the solution of Eq. (121) with $f^n(0) = P^n f_0$, $f_0 \in B_2^+$, $f_0 \ln f_0 \in B$. Then the sequence $\{I^n f^n\}$ contains a subsequence which converges weakly in B_T $\forall T \geqslant 0$.

PROOF. The result follows from Lemma 4 and the following criterion for weak compactness derived in Refs. 14(a),(b).

A sequence $\{f^n\}$ of non-negative functions satisfying the uniform bounds

$$\int_0^T dt \int_{R^3} d^3v \int_\Gamma d^3r (1+v^2) f^n(\vec{r},\vec{v},t) \leqslant K, \quad \int_0^T dt \int_{R^3} d^3v \int_\Gamma d^3r f^n(\vec{r},\vec{v},t) \ln f^n(\vec{r},\vec{v},t) \leqslant K$$

is compact in the L^1 weak topology.

All of the results of this section come from Ref. 38. In that reference, weak compactness arguments are also used to deduce the existence of a solution for generalized collision models, along the lines of those used by Arkeryd.[14(a),(b)] In particular assume $\sigma(V,\mu)$ obeys

$$0 \leqslant \sigma(V,\mu) \leqslant M(1+V^\lambda)$$

for some $0 \leqslant \lambda < 2$, $M > 0$. Define a sequence $\{\sigma_p\}$ of bounded kernels by

$$\sigma_p(V,\mu) = \inf\{\sigma(V,\mu),p\}$$

and write J_p for the collision operator J with kernel σ_p. (For every finite p, $J_p = -\nu_p + L_p$, with ν_p and L_p separately bounded, so the results of Sec. 18 guarantee a solution to Eq. (121) with kernel J_p.)

Then the following result is proved in Ref. 38:

THEOREM XII. Let $f_0 \in B_2^{n+}$, $f_0 \ln f_0 \in B^n$, and f_0 the solution of Eq.(123a) with collision kernel σ_p satisfying $f_0(0) = f_0$. Then $\{f_p(t)\}_{p=1}$ contains a subsequence which converges weakly in B_k^n, $k < 2$. The limit f(t) is continuous, satisfies the bounds $\|f(t)\|_{B^{\bar{n}}} = \|f_0\|_{B^n}$, $\|f(t)\|_2 = \|f_0\|_2$. and obeys Eq. (123a) with (unbounded) collision kernel $\sigma(V.\mu)$.

The proof follows from the compactness criterion used in the proof of Theorem XI, along with the estimates of Lemmas 1 and 3 plus the equicontinuity of the sequence $\{f_0\}_{p=1}^{\infty}$. Note the fact that in Theorem XII, the limiting function is proved to obey the Boltzmann equation; this is not true in Theorem XI.

"Ich habe genug."

J. S. Bach, Cantata No. 82.

ACKNOWLEDGEMENTS

I wish to thank Professor Ivan Kuščer for showing me how to prove the H-theorem (heuristically) for diffusely reflecting boundaries and William Greenberg and Jacek Polawczak for helpful discussions, particularly in the preparation of Chapter V. Miss Deborah Watts did an outstanding job of typing the manuscript under extreme time pressure--without her devoted efforts, this manuscript could not have been prepared on time.

The preparation of this manuscript was supported by the U. S. Department of Energy under Grant No. DE-AS05-80ER10711.

REFERENCES AND FOOTNOTES

1. We adopt the notation and techniques described by James J. Duderstadt and William K. Martin 'Transport Theory" (John Wiley, New York, 1979). Chapter 3. This textbook contains a number of references to the earlier literature.

2. The treatment follows that of Kerson Huang "Statistical Mechanics" (John Wiley, New York, 1963) Chapter 4.

3. I am indebted to Professor Ivan Kuščer for pointing this fact out to me, and for providing me with a sketch of the proof given here.

4. Joel L. Lebowitz and Peter G. Bergmann, Ann. Phys. (N.Y.) 1, 1(1957).

5. H. Grad in Handbuch der Physik, Vol. XII. "Thermodynamics of Gases" (Springer-Verlag, Berlin, 1968).

6. Sometimes, in order to compress notation, we shall use the symbol $(\vec{w})$ to indicate the molecule whose velocity is $\vec{w}$.

7. See, for example, P. F. Zweifel "Reactor Physics" (McGraw-Hill, New York, 1973) Appendix E.

8. H. Goldstein. "Classical Mechanics" (Addison-Wesley, Cambridge, Mass., 1953) page 82.

9. Michael Reed and Barry Simon "Methods of Modern Mathematical Physics-II. Fourier Analysis, Self-Adjointness (Academic Press, New York, 1975) Sec. X.13.

10. Walter Rudin. "Real and Complex Analysis" (McGraw-Hill, New York, 1966) page 21.

11. T. Carleman. "Problemès mathématiques dans la théorie cinétique des gaz," (Almqvist and Wiksells, Uppsala. 1957).

12. E. Wild, Proc. Camb. Phil. Soc. 47, 602(1951).

13. D. Morgenstern, Proc. Nat. Acad. Sci. U.S.A. 40, 719(1954).

14. Lief Arkeryd (a). Arch. Rat. Mech. and Anal. 95, 1(1972);(b)17(1972). c.f. also (c) "Intermolecular Forces of Infinite Range and the Boltzmann Equation," Chalmers University of Technology (Sweden) preprint (1970).

15. A. Ja. Povzner, Mat. Sbornik 58, 65(1962).

16. G. Di Blasio, Boll. U.M.I. 8, 127(1973); Comm. Math. Phys. 38, 331(1974).

17. Shmuel Kaniel and Marvin Shinbrot, Comm. Math. Phys. 58, 65(1978).

18. Alexander Glikson, Arch. Rat. Mech. Anal. 45, 35(1972); Bull. Australian Math. Soc. 16, 321(1977).

19. D. Morgenstern, J. Rat. Anal. 4, 533(1955).

20. Barry Simon "The $P(\phi)_2$ Euclidean (Quantum) Field Theory" (Princeton Univ. Press, Princeton, N.J., 1974).

21. Michael Reed and Barry Simon "Methods of Modern Mathematical Physics--I; Functional Analysis" (Academic Press, New York, 1972) p. 151.

22. The proof of Lemma 2 requires that $R(t_1)R(t_2) = R(t_2)R(t_1)$, otherwise, the "time-ordered" exponential must be used. cf. Chapter VI for a situation in which this is necessary.

23. So we are working, initially, in $L^1(R^3 \times R^3)$.

24. H. Grad in "Applications of Nonlinear Partial Differential Equations in Mathematical Physics" (Amer. Math. Soc., Providence, R.I., 1965) p. 154.

25. cf. Ref. 5, p. 237.

26. cf. Ref. 24, Appendix.

27. cf. Ref. 2, p. 131.

28. H. Grad, Comm. Pure Appl. Math. XVIII, 315(1965).

29. H. Grad, Phys. Fluids 6, 147(1963).

30. Y. Shizuta and K. Asano, Proc. Japan Acad. 53, 3(1977).

31. T. Kato "Perturbation Theory for Linear Operators" (Springer-Verlag, New York, 1966) Chapter IX.

32. T. Nishida and K. Imai, Publ. RIMS Kyoto Univ. 229(1976).

33. J. P. Guraud, Coll. Int. C.N.R.S. No. 236.

34. Y. Shizuta, "On the Classical Solutions of the Boltzmann Equation." Preprint.

35. S. Ukai, Proc. Japan Acad. 50, 179(1974); C.R. Acad. Sci. Paris 282, A-317(1976).

36. C. Cercignani, W. Greenberg and P. F. Zweifel, J. Stat. Phys. 20, 449(1979).

37. Herbert Spohn, J. Stat. Phys. 20, 463(1979).

38. W. Greenberg, J. Voigt and P. F. Zweifel, J. Stat. Phys. 21, 649(1979).

38a. cf. Sec. 12, Lemma 2. Morgenstern's operator $R(t;f)$ is a multiplicative operator, and thus the evolution is given simply by

$$\int_0^t R(t;f(\tau))d\tau .$$

In our case, $T(n;t_1,t_2)$ is generated by $A + \nu(n)$ and is hence a non-diagonal matrix, which is why we find it necessary to introduce "time-ordering." Somehow, Morgenstern has been able to circumvent this difficulty by working along the free trajectories.

39. See Ref. 31, pp. 487 ff.

40. J. P. Ginzburg, Am. Math. Soc. Translations, Series 2, 96, 189(1970).

41. Ref. 21, Sec. VIII.8.

42. J. Voigt, T.T.S.P. 8, 17(1979).

43. M. Moreau, J. Math. hys. 19, 2494(1978).

44. J. Voigt "H-theorem for Boltzmann Type Equations." Preprint, Laboratory for Transport Theory and Mathematical Physics, Va. Polytech. Inst. (1979).

45. H. L. Royden "Real Analysis" (MacMillan, London, 1968) p. 110.

APPENDIX. GENERALIZED EXISTENCE PROOFS AND THE H-THEOREM

In Sec. 3, we presented a heuristic proof of the H-theorem, which ignores the technical questions of existence of the integrals, and similar subtleties In this Appendix we discuss briefly some of the procedures which must be followed in order to convert the heuristic proof into a rigorous proof.

First, it is certainly clear that no H-theorem can be proved for scattering models for which global existence cannot be proved. That is, it is meaningless to talk about n(t)ln n(t) unless it is known that n(t) exists. Thus, one could visualize proving H-theorems for the spatially homogeneous Boltzmann equation (Chapter II), the mollified equation (Chapter IV), the equations describing systems close to equilibrium (Chapter V) or the equation set on a spatial lattice (Chapter VI). It is even conceivable, to think of a "local" H-theorem for the situations in which local existence can be proved (Chapter III).

In any event, we observe the close connection between existence proofs and the H-theorem. In this Appendix we present both a generalized existence proof and a generalized H-theorem, proved by Voigt.[44]

H-theorems have been proved for the spatial homogeneous case by Carleman,[11] Morgenstern[13] and Arkeryd,[14(b)] and for systems near equilibrium by Guiraud.[32] It seems that an H-theorem could have been proved for the mollified Boltzmann equation[13,15] but in fact this was not done.

It would seem fruitful to study the possibility of proving "local" H-theorems as mentioned above, since the establishment of a trend towards equilibrium on a small time interval $[0,t_0]$ should provide help in the attempt to prove existence on the next time interval. However, as far as we know, this has not been done even though the H-theorem crucially enters the proof of existence of the lattice limit[38] and the non-Maxwellian collision model.[14(a)(b),38]

The work of Voigt[44] we describe here involves both existence and the proof of an H-theorem for an abstract Boltzmann equation which includes as special cases the spatially homogeneous equation, the mollified equation and the lattice equation. We briefly sketch Voigt's procedures in the remainder of this Appendix, omitting proofs. Voigt considers only mild solutions; the question of differentiability and hence existence of stronger classical solutions never enters.

Consider, then, the Boltzmann-type equation

$$\frac{du}{dt} = Su + J(u,u), \; u(0) = \phi. \qquad \text{(A-1)}$$

Here S is the generator of the strongly continuous semi-group U_t, $t \geq 0$. (Of course, $S = -\vec{v}\cdot\vec{\nabla}$ for the Boltzmann equation; the treatment here is somewhat more general. Incidentally, U_t is supposed to include boundary conditions.) We seek solutions of u in $L^1(u)$ for a measure space (M,Σ,ν); typically, M = R^3 (spatially homogeneous case), $\Gamma_n \times R^3$ (lattice approximations) etc. We suppose further that J is defined and continuous on $L^1(\mu)$. This is precisely the property which the collision term does not have (cf. Sec. 8), except for the special models we have mentioned earlier. Further we assume J can be decomposed as J = Q - P.

Voigt's analysis begins with a general existence theorem for the mild form of Eq. (A-1) supposing certain conditions are obeyed. These are

(U1) $U_t\phi \geq 0 \; \forall\phi\in L^{1+}(\mu)$, $t \geq 0$, $\|U_t\| \leq e^{mt}$, $t \geq 0$ for some $m \in R$.

(P1) P is locally Lipshitz continuous, uniformly on bounded sets. Further, $\exists R: L^1(\mu) \to L^\infty(\mu)$, linear and continuous, $R(L^{1+}(\mu)) \subseteq L^{\infty+}(\mu)$ s.t. $P(\phi) \leq R(\phi)\phi \; \forall\phi \in L^{1+}(\mu)$.

(Q1) Q is locally Lipschitz continuous, uniformly on bounded sets, and monotone on $L^{1+}(\mu)$ (i.e. $0 \leq Q(\phi) \leq Q(\psi)$ for $0 \leq \phi \leq \psi$).

(J1) $\int J(\phi)(\omega)d\mu \leq m'\|\phi\| \; \forall\phi \in L^{1+}(\mu)$.

Then

THEOREM A-1. Let U_t, P, Q satisfy (U1), (P1), (Q1), (J1). Let $\phi \in L^{1+}(\mu)$. Then $\exists!$ mild solution $f:[0,\infty]\to L^1(\mu)$ of Eq. (A-1). Further, the dependence of the solution on the initial datum is continuous, $f(t) \geq 0 \; \forall t \geq 0$. In addition, the function $[0,\infty] \ni t \to e^{-(m+m')t}\|f(t)\|$ is non-increasing. If in addition

$(\overline{U1})$ $\|U(t)\phi\| = \|\phi\|$, $\phi \in L^{1+}(\mu)$

$(\overline{J1})$ $\int J(\phi)(\omega)d\mu = 0$, $\phi \in L^{1+}(\mu)$

then $\|f(t)\| = \|\phi\| \; \forall t \geq 0$.

Now we introduce the "energy norm." Specifically, let $\alpha: M \to [1,\infty]$ be measurable. $L^1(\mu) \supseteq L^1(\alpha\mu) = \{f \in L^1(\mu) \mid \alpha f \in L^1(\mu)\}$ with norm $\|f\|_\alpha = \|\alpha f\|_{L^1}$. Introduce the properties of U, P and Q with respect to $L^1(\alpha\mu)$ similar to those they possess with respect to $L^1(\mu)$.

(U2) The restriction $(U_t^\infty;\ t \geqslant 0)$ of $(U_t; t \geqslant 0)$ to $L^1(\alpha\mu)$ is a strongly continuous semigroup on $L^1(\alpha\mu)$ and $\|U_t^\infty\|_{L^1(\alpha\mu)} \leqslant e^{m_\alpha t}$, $t \geqslant 0$ for some $m_\alpha \in R$. We denote the generator of U_t^∞ by S_α. Since $L^1(\alpha\mu)$ is continuously imbedded in $L^1(\mu)$, $S_\alpha \subseteq S$.

($\overline{U2}$) In addition to (U2), $\|U(t)\phi\|_\alpha = \|\phi\|_\alpha$, $\phi \in L^{1+}(\alpha\mu)$, $t \geqslant 0$.

(P2) [resp. (Q2)] $P(L^1(\alpha\mu))$ [resp. $Q(L^1(\alpha\mu))$] $\subseteq L^1(\alpha\mu)$ and P(resp. Q) is locally Lipschitz continuous uniformly on bounded sets of $L^1(\alpha\mu)$.

(J2) $\int \alpha(\omega) J(\phi)(\omega) d\mu \leqslant m_\alpha' \|\phi\|_\alpha \quad \forall \phi \in L^{1+}(\alpha\mu)$.

($\overline{J2}$) $\int \alpha(\omega) J(\phi)(\omega) d\mu = 0 \quad \forall \phi \in L^{1+}(\alpha\mu)$.

Then

THEOREM A-2. Let (U1), (U2), (P1), (P2), (Q1), (Q2), (J1), (J2) be satisfied and $\phi \in L^{1+}(\alpha\mu)$. Then the mild solution described in Theorem A-1 is a mild solution in $L^1(\alpha\mu)$ and $e^{-(m_\alpha + m_\alpha')t} \|f(t)\|_\alpha$ is a non-increasing function of t. If ($\overline{U2}$), ($\overline{J2}$) are satisfied, then $\|f(t)\|_\alpha = \|\phi\|_\alpha \ \forall t \geqslant 0$.

We now turn to the question of the H-theorem. For functions $\phi \in L^{1+}(\mu)$ s.t. $\phi \ln \phi \in L^1(\mu)$ define as usual

$$H(\phi) = \int \phi \ln \phi \, d\mu. \tag{A-2}$$

Understanding that $0 \cdot \ln \phi = 0$, then function $[0,\infty] \ni y \to y \ln y$ is continuous.

Referring to the heuristic (or "formal") proof of the H-theorem as given in Sec. 3, we recall that we expressed

$$\frac{dH(f(t))}{dt} = \int (Sf(t))(\ln f(t) + 1) d\mu + \int J(f,f)(t)(\ln f(t) + 1) d\mu \tag{A-3}$$

and argued that both terms on the right-hand side were negative, the first by analyzing the boundary conditions in detail (using the Stückelberger-Pauli identity) and the second through making four changes of variable of integration. The main problem in this proof is due to the fact that $H(\phi)$ is not continuous on $L^{1+}(\mu)$ or $L^{1+}(\alpha\mu)$. A second problem is that we may be dealing with mild solutions, so the steps leading to Eq. (A-1), involving differentiation of f with respect to t may not be justifiable. To circumvent these difficulties, it is necessary to impose further conditions.

(U3) $\exists$ constants M_∞, m_∞ s.t. $\|U_t\phi\|_\alpha \le M_\infty e^{m_\infty t} \|\phi\|_\infty$, $t \ge 0$ $\forall\phi \in L^1(\mu) \cap L^\infty(\mu)$. (This would be trivially true if U_t were a strongly continuous semigroup on $L^\infty(\mu)$ which, in general, it is not.)

(α1) Let α: $M \to [1,\infty)$ be measurable with $e^{-\alpha} \in L^1(\alpha\mu)$.

(U4) $\exists$ constants $M_0 > 0$, m_0 s.t. $U_t e^{-\alpha} > M_0 e^{m_0 t} e^{-\alpha}$, $t \ge 0$.

(U5) If $\phi \in L^{1+}(\alpha\mu)$ s.t. $\phi \ln \phi \in L^1(\mu)$, then $(U_t\phi)\ln(U_t\phi) \in L^1(\mu)$, $t \ge 0$, and $[0,\infty) \ni t \to H(U_t\phi)$ is non-increasing.

We observe that (U5) implies an H-theorem for the free motion (including boundary conditions, which are, we recall, supposed to be built into U_t).

(P3) $P(\phi) \ge 0$, $\phi \in L^{1+}(\mu)$.

(J3) For $\phi \in L^{1+}(\alpha\mu)$ s.t. $\phi \ln \phi \in L^1(\mu)$ and $\varepsilon e^{-\alpha} \le \phi \le K$ for some ε, $K > 0$, then $\int J(\phi)(\ln \phi + 1)d\mu \le 0$.

We finally can state

<u>THEOREM A-3, H-THEOREM</u>. Let (U1)-(U5), (P1)-(P3), (Q1), (Q2), (J1)-(J3), (α1) be satisfied. Let $\phi \ln \phi \in L^1(\mu)$ and f: $[0,\infty) \to L^{1+}(\alpha\mu)$ be the mild solution of

$$\frac{du}{dt} = S(u) + J(u,u), \quad u(0) = \phi$$

described in Theorem A-2. Then $f(t)\ln f(t) \in L^1(\mu)$ $\forall t \ge 0$ and $H(f(t))$ is a non-increasing function of t.

The proof of this theorem is long and technical, and we have no intention of reproducing it here. We observe, however, how many technical conditions are necessary in order to improve the formal proof of the H-theorem to a rigorous proof. One might ask whether this plethora of conditions can ever be verified in a practical case.

The answer is, in fact, that it can, and Voigt shows[44] that both the lattice approximation (Chapter VI) and the mollified form of the Boltzmann equation

(Chapter IV) are models which satisfy the conditions for the H-theorem.

In particular, consider the lattice model. Then ($\overline{U1}$) and ($\overline{U2}$) follow from Lemma 1, Sec. 19. The function α was taken to be

$$\alpha(\vec{r},\vec{v}) = 1+v^2$$

which implies that (α1) is satisfied. (U3) and (U4) follow from Corollary 1 to Lemma 1, Sec. 18, (U5) follows from Lemma 3, Sec. 19. (P1)-(P3) and (Q1), (Q2) follow from the assumed separate boundedness of P(=ν in Chapter VI) and Q(=L in Chapter VI). ($\overline{J1}$) and ($\overline{J2}$) follow from the fact that 1 and v^2 are collision invariants and (J3) for the standard inequality $(y-y')(\ln y-\ln y') \geqslant 0$, $y,y' > 0$.

PRELIMINARY RESULTS ON THE NON-EXISTENCE OF SOLUTIONS FOR A HALF SPACE BOLTZMANN COLLISION MODEL WITH THREE DEGREES OF FREEDOM

M. D. Arthur

Abteilung für Mathematische Physik
Universität Ulm

ABSTRACT.

A two component model of the linearized Boltzmann equation in a half-space incorporating three degrees of freedom is studied. The linearized collision term is taken to be a summation over a suitable combination of the collision invariants. Preliminary results concerning exponents of compensating factors involved in the Wiener-Hopf factorization of the dispersion matrix are presented. Comparison is made to the case of the linearized Boltzmann equation in a half-space incorporating only one degree of freedom.

I. INTRODUCTION

In an earlier seminar[1] we heard a discussion on the breakdown of solution to the non-linear Boltzmann equation,

$$\frac{\partial f}{\partial t} + \vec{\xi}" \cdot \frac{\partial f}{\partial \vec{x}"} = \Omega\ (f,f)\ , \tag{1}$$

modeling a Knudsen boundary layer for Mach numbers exceeding unity. In this paper we are thinking of the same physical problem, however, we consider the steady state linearized Boltzmann equation for $\tilde{h}$, the deviation from a downstream Maxwellian distribution. In particular, we set

$$f = F(1+\tilde{h}) \tag{2 a}$$

$$F = \left(\frac{\rho_\infty}{2\pi RT_\infty}\right)^{1/2} \exp\left[-\frac{1}{2RT_\infty}(\vec{\xi}'' - \vec{v}_\infty)\right] \quad (2\text{ b})$$

where $\vec{\xi}''$ is the molecular velocity and ρ_∞, T_∞ and $\vec{v}_\infty$ are the downstream temperature, density and drift velocity respectively.

The steady state state linearized version of Eq. (1) for one spatial dimension is given by

$$\xi_1'' \frac{\partial \tilde{h}}{\partial x''} = L\,\tilde{h} \quad (3)$$

where L is obtained from Q by omitting terms quadratic in $\tilde{h}$. For the special case of "quasi" Maxwell molecules (i.e. inverse fifth power force law with angular cut-off) the eigenvalue problem for L

$$L\tilde{\psi}_i = \lambda_i \tilde{\psi}_i \,, \quad (4)$$

has been solved exactly[2] and the eigenfunctions are given by products of generalized Laguerre polynomials times spherical harmonics. The operator L for this model may conveniently be written as

$$L\tilde{h} = \nu\left[-\tilde{h} + \sum_{i=0}^{4} \tilde{\psi}_i(\tilde{\psi}_i,\tilde{h})_3 + \sum_{i=5}^{\infty} (1+\lambda_i)\tilde{\psi}_i(\tilde{\psi}_i,\tilde{h})_3\right], \quad (5\text{ a})$$

$$(\tilde{\phi},\tilde{h})_3 = \int \tilde{\phi}\,\tilde{h}\,F\,d^3\xi'' \,, \quad (5\text{ b})$$

where the first five $\tilde{\psi}_i$ are suitable combinations of the well known collision invariants (usually denoted by ψ_α) and are the eigenfunctions corresponding to zero eigenvalue. Here ν is a constant on the order of an inverse collision time.

The rest of this paper concerns itself with the case for which the infinite summation in Eq. (5 a) is omitted. This case can be arrived at without specific reference to the "quasi" Maxwell molecule model[3], however, we have chosen this line of development in order to indicate that the model we consider may be generalized along the lines of Ref. 4.

We introduce the following change to dimensionless variables,

$$x = (\nu / \sqrt{2RT_\infty})\; x'' \,, \quad (6\text{ a})$$

$$\vec{\xi} = (\vec{\xi}'' - \vec{v}_\infty) \; / \; \sqrt{2RT_\infty} \quad , \tag{6 b}$$

$$h(x,\vec{\xi}) = \tilde{h}(x'',\vec{\xi}'') \quad , \tag{6 c}$$

so that Eq. (3) becomes

$$(\xi_1 + u) \frac{\partial h}{\partial x} = \sum_{i=0}^{4} \psi_i(\vec{\xi}) \; (\psi_i, h)* - h \; , \tag{7 a}$$

$$(\phi, h)* = \pi^{-3/2} \int \phi \, h \, e^{-\xi^2} d^3\xi \; . \tag{7 b}$$

Here, $u = v_\infty / \sqrt{2RT_\infty}$ is the speed ratio. There is a critical value of u which corresponds to Mach number unity and is given by $[(j+2) / 2j]^{1/2}$ for molecules with j degrees of freedom. Although Eq. (7 a) contains only one spatial coordinate, the presence of three "velocity" components will allow us to develop a model with three degrees of freedom.

In a recent paper[5], a one "velocity" component (one degree of freedom) version of Eq. (7 a) was studied. It was shown that there exists a unique solution to the half-space problem which admits at most two additional conditions on the incoming distribution (corresponding to specification of downstream temperature and density) only for speed ratios u such that $u < \sqrt{3/2}$. Thus, in general, there does not exist a unique, physically acceptable solution to the steady state, linearized Boltzmann equation in a half-space for a supersonic flow situation. The importance of this result is due to the fact that physical and numerical evidence show this result for the linearized case also holds for the nonlinearized case, with perhaps the only change of $u < \sqrt{3/2}$ to $u \le \sqrt{3/2}$.

The discussion in Ref. 1 provides us with the physical and numerical evidence to show non-existence of a solution to the non-linear Boltzmann equation for molecules with three degrees of freedom in a supersonic flow situation. Therefore, it is of interest to generalize the results of Ref. 5 to molecules with three degrees of freedom and thus provide a rigorous proof for the linearized version of the problem discussed in Ref. 1.

At present, the generalization of the results in Ref. 5 is not complete. However, the preliminary results to be presented here are strikingly similar. In particular, since we are concerned with a half space problem, it is necessary to construct the Wiener-Hopf factorization of the dispersion "function" (called Ω) for the problem.

Generally, two functions X and Y are said to be a Wiener-Hopf factorization of Ω provided i) $\Omega = XY$ and ii) X is analytic in a given half plane of the complex plane and Y is analytic in its complement. It is important to note that for any given Wiener-Hopf factorization of Ω, we could always multiply X by a suitably chosen function g (e.g. any polynomial having its zeros in the closure of the X function's half-plane of analyticity) and multiply Y by $1/g$ so that the new functions also constitute a Wiener-Hopf factorization of Ω. In Ref. 5, the function $g(\nu) = (\nu + u)^{-\beta}$, $\nu \in \mathbb{C}$, β any integer, was chosen. However, it was shown that not all values of β are permissible; i.e.

$$-3 \le \beta \le -1 \qquad \text{for} \quad u < \sqrt{3/2} \tag{8 a}$$

and

$$-3 \le \beta \le 0 \qquad \text{for} \quad u > \sqrt{3/2} \tag{8 b}$$

A subsequent asymptotic analysis for each permissible choice of β and u revealed that, in general, only $\beta = -1$ for $u < \sqrt{3/2}$ leads to a unique, physically acceptable solution. We note that from (8 a) and (8 b), the supersonic case has one more permissible value of β than the subsonic case. It is this additional value of β which marks the supersonic case.

The preliminary result to be presented here is that the supersonic case for the linearized Boltzmann equation for molecules with three degrees of freedom is marked in a manner analogous to (8 a) and (8 b). In Sec. II, we present a more manageable form of Eq. (7 a) having a reduced explicit dependence on three velocities at the expense of an increased number of equations to consider[6], nevertheless maintaining the overall three degrees of freedom character of the original model. The coupled equations are written in a compact, operator notation so that a formal solution can be written in terms of the resolvent of the relevant transport operator. The resolvent of the transport operator is calculated in Sec. III, and the Wiener-Hopf factorization of the dispersion matrix which appears is discussed. Sec. IV contains the Ω function argument which is the essence of the Larsen-Habetler[7,8,9] method for solving transport equations. The preliminary result (i.e. the analogue of (8 a) and (8 b) for molecules with three degrees of freedom) arises from this Ω function argument. Some concluding remarks are found in Sec. V.

II. REDUCED DESCRIPTION AND THE TRANSPORT OPERATOR

There are six dependent variables in Eq. (7 a), thus any direct solution is likely to be highly complicated. One method of reducing the number of dependent variables would be to integrate two of the velocity variables, say ξ_1 and ξ_2. However, such a procedure could result in the loss of the three degrees of freedom character for Eq. (7 a). Another method, developed in Ref. 6, which preserves the three degrees of freedom character of Eq. (7 a) is to regard h as a function in a Hilbert space, H, of functions of ξ_2 and ξ_3 with x and ξ_1 as fixed parameters. We then choose a set of basis functions for a finite dimensional subspace of H and project h onto each basis function, each time obtaining a function of x and ξ_1. There is, of course, a part of h that is orthogonal to the finite dimensional subspace of H spanned by the chosen set of basis functions and it too is a function of x and ξ_1.

We define,

$$\phi_0 = 1 \quad , \tag{9 a}$$

$$\phi_1 = \xi_2^2 + \xi_3^2 - 1 \, , \tag{9 b}$$

$$\phi_2 = \xi_2 \quad , \tag{9 c}$$

and

$$\phi_3 = \xi_3 \quad ,$$

so that h may be written as

$$h(x,\vec{\xi}) = Y_0(x,\xi_1)\,\phi_0 + Y_1(x,\xi_1)\,\phi_1 + 2\,Y_2(x,\xi_1)\,\phi_2 + 2\,Y_3(x,\xi_1)\,\phi_3 + Y_4(x,\xi_1,\xi_2,\xi_3) \, . \tag{10}$$

We now multiply Eq.(7 a) by $\phi_j \exp[-\xi_2^2 - \xi_3^2]$; $j = 0,1$, and integrate with respect to ξ_2 and ξ_3. The result is a system of two coupled equations for Y_0 and Y_1 given by

$$(\xi_1 + u)\,\frac{\partial Y_0(x,\xi_1)}{\partial x} + Y_0(x,\xi_1) = \pi^{-1/2}\int_{-\infty}^{\infty}[1 + 2\xi_1\xi_1' + \frac{2}{3}(\xi_1^2 - \frac{1}{2})(\xi_1'^2 - \frac{1}{2})]\,d\xi_1'$$

$$+ \pi^{-1/2}\,\frac{2}{3}(\xi_1^2 - \frac{1}{2})\int_{-\infty}^{\infty} Y_1(x,\xi_1')\,\exp(-\xi_1')\,d\xi_1' \, , \tag{11 a}$$

and

$$(\xi_1 + u) \frac{\partial Y_1(x,\xi_1)}{\partial x} + Y_1(x,\xi_1) = \frac{2}{3} \pi^{-1/2} [\int_{-\infty}^{\infty} (\xi_1'^2 - \frac{1}{2}) Y_0(x,\xi_1') \exp(-\xi_1'^2) \, d\xi_1'$$

$$+ \int_{-\infty}^{\infty} Y_1(x,\xi_1') \exp(-\xi_1'^2) \, d\xi_1']. \tag{11 b}$$

Of course, additional equations for Y_2 and Y_3 may be obtained in the same manner and there is also an equation for Y_4, however, none of the equations for Y_2, Y_3 and Y_4 involve Y_0 or Y_1 (see Ref. 6). Equations (11 a) and (11 b) describe heat transfer effects while the equations for Y_2 and Y_3 describe shear effects and we shall only be concerned with the former.

It will be easier to perform manipulations if we make use of some matrix notation. We define

$$\psi(x,\xi) = \begin{bmatrix} Y_0(x,\xi) \\ \\ Y_1(x,\xi) \end{bmatrix} , \tag{12}$$

where we have dropped the subscript from the x-component of the velocity variable since now only one velocity component will appear explicitly. The coupled system of equations now become

$$\frac{\partial \psi(x,\xi)}{\partial x} + \underset{\approx}{K}[\psi](x,\xi) = 0 , \tag{13 a}$$

where the transport operator $\underset{\approx}{K}$ is given by

$$\underset{\approx}{K}[\psi](x,\xi) = \frac{1}{\xi+u} [\underset{\approx}{I}\psi(x,\xi) - \pi^{-1/2} \int_{-\infty}^{\infty} \underset{\approx}{M}(\xi,-\xi')\psi(x,\xi') e^{-\xi'^2} d\xi'] \tag{13 b}$$

$$\underset{\approx}{M}(\xi,-\xi') = \begin{bmatrix} 1 + 2\xi\xi' + \frac{2}{3}(\xi^2 - \frac{1}{2})(\xi'^2 - \frac{1}{2} & \frac{2}{3}(\xi^2 - \frac{1}{2}) \\ \\ \frac{2}{3}(\xi'^2 - \frac{1}{2}) & \frac{2}{3} \end{bmatrix} , \tag{13 c}$$

and $\underset{\approx}{I}$ is the 2×2 identity matrix.

We shall seek a solution to Eq. (13 a) in a half-space, $x \geq 0$ which satisfies the boundary conditions

$$\psi(x,\xi) \to 0 \qquad \text{as} \quad x \to +\infty \quad , \tag{14 a}$$

and

$$\psi(0,\xi) = \hat{g}(\xi) \qquad \text{for} \quad \xi > -u \quad , \tag{14 b}$$

where $\hat{g}$ is a given function. Boundary condition (14 b) is in a somewhat non-standard form. Perhaps a more familiar form is

$$\psi(0,\xi) = \alpha\psi(0,-\xi) + (1-\alpha)\int_{-\infty}^{\infty} \sigma(\xi,\xi' \to \xi)\psi(0,\xi')\, d\xi' \ , \qquad \xi < -u \tag{15}$$

where σ is in some sense a probability measure for diffuse scattering at the origin and α is an accomodation coefficient ($\alpha = 1$ corresponds to specular reflection). Boundary conditions like Eq. (15) have also been discussed more abstractly in terms of a stochastic kernel[10]. However, we shall take the point of view that a complete solution to Eq. (13 a) must specify the outcoming distribution ($\psi(0,\xi)$, $\xi < -u$) in terms of the incoming distribution ($\psi(0,\xi)$, $\xi > -u$) and that with such a specification, a boundary condition of the form given by Eq. (15) may be regarded as an integral equation for $\hat{g}$.

With some abuse of notation and keeping in mind the distinction between incoming distribution and outcoming distribution, we define,

$$\psi_0(\xi) \equiv \psi(0,\xi) \quad . \tag{16}$$

Then the formal solution to Eq. (13 a) may be written as

$$\psi(x,\xi) = -\frac{1}{2\pi i}\int_{\Gamma} (\underset{\approx}{K} - z\underset{\approx}{I})^{-1}\,[\psi_0](\xi)\exp(-xz)\,dz \ , \tag{17}$$

where Γ is an appropriate contour in the complex z-plane, to the "right" of which lie all the singularities of the integrand (or equivalently, one which surrounds the spectrum of $\underset{\approx}{K}$). The problem with the solution given by Eq. (17) is that in its present form, the resolvent of $\underset{\approx}{K}$, $(\underset{\approx}{K} - z\underset{\approx}{I})^{-1}$, i) has a simple pole singularity at $z = 0$ which would give rise to constant, non-decaying modes violating boundary condition (14 a) and ii) is discontinuous across the negative real axis, violating (14 a) and also making it impossible to define an appropriate contour Γ. Both of these problems can be eliminated by the requirement that the resolvent of $\underset{\approx}{K}$ be analytic for $\mathrm{Re}\, z \le 0$. The analyticity requirement at $z = 0$ is actually superfluous in our case since application of it leads to the condition

$$\int_{-\infty}^{\infty} \xi^j Y_0(x,\xi) e^{-\xi^2} d\xi = (-u)^j \int_{-\infty}^{\infty} Y_0(x,\xi) e^{-\xi^2} d\xi \quad , \quad j = 1,2 \tag{18}$$

which is merely the statement of momentum and energy conservation inherent in the original, linearized Boltzmann equation from which Eq. (13 a) is derived. An additional difficulty with the solution given by Eq. (17) is that it does not specify the outcoming distribution in terms of the incoming distribution and hence our point of view concerning boundary conditions discussed in the previous paragraph is in jeopardy. However, it will be seen that the analyticity requirement along with the Q function argument of Larsen and Habetler leads to the necessary specification.

We remark hear that our ideal goal is to show that something is irreparably wrong with the solution given by Eq. (17) for supersonic downstream conditions. We are using the fact that Eq. (13 a) has a unique solution given by Eq. (17) if and only if $\underset{\approx}{K}$ is the generator of a strongly continuous semigroup[11]. The only semigroup property we will need in order to arrive (or more properly - partially arrive) at our ideal goal is that the resolvent set of $\underset{\approx}{K}$ contain some unbounded portion of the real axis. The question as to which portion of the real axis is contained in the resolvent set of $\underset{\approx}{K}$ is, in our case, directly related to boundary condition (14 a) and the answer turns out to be the entire negative real axis. Hence, our analyticity requirement is merely another version of one part of the semigroup requirement. We shall tacitly assume that there is some Banach space (perhaps not the most ideal) on which $\underset{\approx}{K}$ satisfies the norm condition for a prospective generator of a semigroup.

The irreparable difficulty with the solution given by Eq. (17) arises from the physical consideration that if ψ_0 is completely specified, then one has to leave ρ_∞ and T_∞ unspecified and assign v_∞ (related to the speed ratio u). Hence, in general, only two independent conditions on ψ_0 may be permitted, thus corresponding to specification of ρ_∞ and T_∞ (regarding u as a parameter).

In order to apply our analyticity requirement, we will need to calculate the resolvent of $\underset{\approx}{K}$ explicitly. This calculation is greatly simplified by the observation that in the definition of $\underset{\approx}{K}$, the integral containing $\underset{\approx}{M}$ only involves the first column of $\underset{\approx}{M}$ and, using Eq. (18), it is easy to show that

$$\underset{\approx}{K}[\psi]\,(x,\xi) = \frac{1}{\xi+u}\left[\underset{\approx}{I}\psi(x,\xi) - \pi^{-1/2}\,\underset{\approx}{M}(\xi,u)\int_{-\infty}^{\infty}\psi(x,\xi')e^{-\xi'^2}d\xi'\right] . \tag{19}$$

III. RESOLVENT OF $\underset{\approx}{K}$ AND THE DISPERSION MATRIX

The complex variable z in Eq. (17) is the Laplace transform variable. It is more convenient to use the Case separation variable ν defined by,

$$z = \frac{1}{\nu+u} \quad . \tag{20}$$

We now write,

$$(\underset{\approx}{K} - \frac{1}{\nu+u}\underset{\approx}{I}) [\tilde{g}] (\xi) \equiv \psi_0(\xi) \quad , \tag{21}$$

and solve for $\tilde{g}$ in terms of ψ_0. We first obtain

$$\tilde{g}(\xi) = \frac{\nu+u}{\nu-\xi} [(\xi+u)\ \psi_0(\xi) + \pi^{-1/2} \underset{\approx}{M}(\xi,u) \int_{-\infty}^{\infty} g(\xi')\ e^{-\xi'^2} d\xi'] \ . \tag{22}$$

Multiplying Eq. (22) by $e^{-\xi^2}$ and integrating with respect to ξ yields the integral term involving $\tilde{g}$,

$$\int_{-\infty}^{\infty} \tilde{g}(\xi')\ e^{-\xi'^2} d\xi' = \underset{\approx}{\Omega}^{-1}(\nu)\ (\nu+u) \int_{-\infty}^{\infty} \frac{\xi'+u}{\nu-\xi'} \psi_0(\xi')\ e^{-\xi'^2} d\xi' \quad , \tag{23 a}$$

where

$$\underset{\approx}{\Omega}(\nu) = \underset{\approx}{I} - \pi^{-1/2} (\nu+u) \int_{-\infty}^{\infty} \frac{1}{\nu-\xi} \underset{\approx}{M}(\xi,u)\ e^{-\xi^2} d\xi \ . \tag{23 b}$$

Combining Eqs (22) and (23 a) we obtain an explicit expression for the resolvent of $\underset{\approx}{K}$ given by

$$(\underset{\approx}{K} - \frac{1}{\nu+u}\underset{\approx}{I})^{-1} [\psi_0](\xi) = \frac{\nu+u}{\nu-\xi} [(\xi+u)\ \psi_0(\xi) + \pi^{-1/2} (\nu+u)\ \underset{\approx}{M}(\xi,u)\ \underset{\approx}{\Omega}^{-1}(\nu)$$
$$\int_{-\infty}^{\infty} \frac{\xi'+u}{\nu-\xi'} \psi_0(\xi')\ e^{-\xi'^2} d\xi'] \ . \tag{24}$$

The matrix $\underset{\approx}{\Omega}$ is the dispersion matrix for our problem and it depends on both ν and u (we have suppressed the u dependence on the right hand side of Eq. (23 b)). The integral containing $\underset{\approx}{M}$ in the definition of $\underset{\approx}{\Omega}$ involves the Hilbert transform of the three functions $\xi^n e^{-\xi^2}$, $n = 0,1,2$. These three Hilbert transforms may be written as

$$\pi^{-1/2}\int_{-\infty}^{\infty}\frac{\xi^n e^{-\xi^2}}{\xi-\nu}\,d\xi = (1-\delta_{n0})\,\nu^{n-1}+\nu^n P(\nu), \qquad n=0,1,2\ , \tag{25}$$

where $P(\nu)$ is related to the plasma dispersion function $P(s)$, in the sense that as $\nu \to s \in \mathbb{R}$, $P(\nu) \to P(s) \pm i\,\pi^{1/2} e^{-s^2}$ where the $+$ $(-)$ sign is taken when ν approaches s from above (below) the real axis. Thus each component of $\underset{\approx}{\Omega}$ (hence, each component of the resolvent of $\underset{\approx}{K}$ given by Eq. (24)) has a jump discontinuity across the entire real axis.

Using Eq. (25), the dispersion matrix $\underset{\approx}{\Omega}$ becomes

$$\underset{\approx}{\Omega}(\nu) = \underset{\approx}{L}(\nu) + (\nu+u)\ P(\nu)\ \underset{\approx}{M}(\nu,u)\ , \tag{26 a}$$

where

$$\underset{\approx}{L}(\nu) = \begin{bmatrix} \frac{2}{3}(u^2-\frac{1}{2})\,\nu^2 + \frac{2}{3}(u^2-\frac{1}{2})\,u\nu - 2(u^2-\frac{1}{2}) & \frac{2}{3}\,\nu(\nu+u) \\ 0 & 1 \end{bmatrix}. \tag{26 b}$$

A more convenient form of $\underset{\approx}{\Omega}$ is obtained from Eq. (26 a) by factoring $\underset{\approx}{M}(\nu,u)/\|\underset{\approx}{M}(\nu,u)\|$ from the right ($\|\underset{\approx}{M}(\nu,u)\| = \frac{2}{3}(1-2u\nu)$ is the determinant of $\underset{\approx}{M}(\nu,u)$), thus

$$\underset{\approx}{\Omega}(\nu) = [\underset{\approx}{N}(\nu) + \frac{2}{3}(1-2u\nu)(\nu+u)P(\nu)\ \underset{\approx}{I}\,]\,\underset{\approx}{M}(\nu,u)\,/\,\|\underset{\approx}{M}(\nu,u)\|\ , \tag{27 a}$$

where the symmetric matrix $\underset{\approx}{N}(\nu)$ is given by

$$\underset{\approx}{N}(\nu) = \frac{2}{3}\begin{bmatrix} -2[\nu u + (u^2-\frac{1}{2})] & -(u^2-\frac{1}{2}) \\ -(u^2-\frac{1}{2}) & (u^2-\frac{1}{2})\nu^2 - 3u\nu - \frac{1}{2}(u^2-\frac{1}{2}) \end{bmatrix}. \tag{27 b}$$

Since $\underset{\approx}{N}$ is symmetric, there exists a matrix $\underset{\approx}{V}$ such that $\underset{\approx}{V}^{-1}\underset{\approx\approx}{NV}$ is diagonal, so that the dispersion matrix may be written as

$$\underset{\approx}{\Omega}(\nu) = \underset{\approx}{V}(\nu,u)\ \underset{\approx}{D}(\nu,u)\ \underset{\approx}{V}^{-1}(\nu,u)\ \underset{\approx}{M}(\nu,u)\,/\,\|\underset{\approx}{M}(\nu,u)\|\ , \tag{28}$$

where D is a diagonal matrix whose diagonal entries consists of the eigenvalues of

$\underset{\approx}{N}$ with $\frac{2}{3}(1-2u\nu)(\nu+u)P(\nu)$ added.

Unfortunately, the eigenvalues of $\underset{\approx}{N}$ involve the square root of a fourth order polynomial in ν. The plasma dispersion function as a function of the complex variable ν already has a branch cut across the entire real axis so that it will be convenient to choose the square root branch cuts in a manner which avoids intersecting the real axis. The diagonalizing matrix $\underset{\approx}{V}$ and its inverse are then analytic everywhere in the complex ν-plane except across the square root branch cuts and the diagonal matrix $\underset{\approx}{D}$ is analytic everywhere in the complex ν-plane except across both the square root branch cuts and the entire real axis. However, the product $\underset{\approx}{V}\,\underset{\approx}{D}\,\underset{\approx}{V}^{-1}$ only has a branch cut across the real axis. Therefore, we seek[12] a Wiener-Hopf factorization of each element of $\underset{\approx}{D}$, say D_1 and D_2, such that

i) $D_1(\nu,u) = X_1(\nu,u)\,Y_1(\nu,u)$ and $D_2(\nu,u) = X_2(\nu,u)\,Y_2(\nu,u)$

where X_1, X_2 are analytic for $\operatorname{Re}\nu < -u$ and Y_1, Y_2 are analytic for $\operatorname{Re}\nu > -u$ [Note: from Eq. (20), $\operatorname{Re} z < 0$ if and only if $\operatorname{Re}\nu < -u$], and

ii) $\hat{\underset{\approx}{X}} = \underset{\approx}{V}\,\underset{\approx}{X}\,\underset{\approx}{V}^{-1}$ and $\hat{\underset{\sim}{Y}} = \underset{\approx}{V}\,\underset{\approx}{Y}\,\underset{\approx}{V}^{-1}$ are analytic in the complex plane except across an unbounded portion of the real axis greater than $-u$ and less than $-u$ respectively,

where the diagonal matrix $\underset{\approx}{X}$ has components X_1, X_2 and the diagonal matrix $\underset{\approx}{Y}$ has components Y_1, Y_2. Then the dispersion matrix $\underset{\approx}{\Omega}$ may be written in the factorized form

$$\underset{\approx}{\Omega}(\nu) = \hat{\underset{\approx}{Y}}(\nu,u)\,\hat{\underset{\approx}{X}}(\nu,u)\,\underset{\approx}{M}(\nu,)\,/\,\|M(\nu,u)\| \quad . \tag{29}$$

Note that the Wiener-Hopf factorization for each of D_1 and D_2 will involve and undetermined integer, say β_1 and β_2 respectively (see the discussion preceeding (8 a) in Sec. I).

IV. APPLYING THE ANALYTICITY REQUIREMENT

Using the factorized form of $\underset{\approx}{\Omega}$, Eq. (24) becomes

$$(\underset{\approx}{K} - \frac{1}{\nu+u}\underset{\approx}{I})^{-1}[\psi_0](\xi) =$$

$$\frac{\nu+u}{\nu-\xi}\,[(\xi+u)\psi_0(\xi) + \pi^{-1/2}(\nu+u)\underset{\approx}{M}(\xi,u)\underset{\approx}{M}^{-1}(\nu,u)\,\|\underset{\approx}{M}(\nu,u)\|$$

$$\cdot\,\hat{\underset{\approx}{X}}^{-1}(\nu,u)\hat{\underset{\approx}{Y}}^{-1}(\nu,u)\int_{-\infty}^{\infty}\frac{\xi'+u}{\nu-\xi'}\,\psi_0(\xi')\,e^{-\xi'^2}\,d\xi']\,. \tag{30}$$

We would like to modify the right hand side of Eq. (30) so that it appears analytic for $\mathrm{Re}\,\nu < -u$. We observe that the term $\frac{1}{\nu-\xi}$ is a simple pole singularity for all real ν and that the term

$$\hat{\underset{\approx}{Y}}^{-1}(\nu,u)\int_{-\infty}^{\infty}\frac{\xi'+u}{\nu-\xi'}\,\psi_0(\xi')\,e^{-\xi'^2}\,d\xi' \tag{31}$$

does not appear to be analytic for $\mathrm{Re}\,\nu < -u$. The remaining terms on the right hand side of Eq. (30) are analytic for $\mathrm{Re}\,\nu < u$.

We now outline the procedure by which the analyticity difficulties just mentioned may be removed.

STEP 1: In order to remove the simple pole singularity at $\nu = \xi < -u$, require the quantity in brackets on the right hand side of Eq. (30) to be zero as $\nu \to \xi < -u$.

STEP 2: Define the vector valued function $\underset{\sim}{\Omega}$ by

$$\underset{\sim}{\Omega}(\nu) = \hat{\underset{\approx}{Y}}^{-1}(\nu,u)\int_{-\infty}^{\infty}\frac{\xi'+u}{\nu-\xi'}\,\psi_0(\xi')e^{-\xi'^2}\,d\xi' - \int_{-u}^{\infty}\hat{\underset{\approx}{Y}}^{-1}(\xi',u)\,\frac{\xi'+u}{\nu-\xi'}\,\psi_0(\xi')e^{-\xi'^2}\,d\xi' \tag{32}$$

and require each component of $\underset{\sim}{\Omega}$ to be bounded at $\nu = -u$ and $\nu = \infty$.

The immediate consequence of STEP 1 is a set of integral equations for the outcoming distribution in terms of the incoming distribution since ν may approach $\xi < -u$ from above or below the real axis,

$$\psi_0(\xi) = \frac{2}{3}\,\pi^{-1/2}\,(1-2u\xi)\,\hat{X}^{-1}(\xi,u)\,\hat{Y}^{-1\pm}(\xi,u)$$

$$[\int_{-\infty}^{\infty}\frac{P}{-}\,\frac{\xi'+u}{\xi'-\xi}\,\psi_0(\xi')e^{-\xi'^2}\,d\xi' \mp i\pi\,(\xi+u)\,\psi_0(\xi)e^{-\xi^2}]\,, \qquad \xi < -u \tag{33}$$

where

$$\hat{\underset{\approx}{Y}}^{-1\pm}(\xi,u) = \lim_{\varepsilon\to 0}\hat{\underset{\approx}{Y}}^{-1}(\xi \pm i\varepsilon,u)\,, \tag{34}$$

and the P under the integral denotes the Cauchy principal value.

The function Q, defined by Eq. (32), is analytic across the real axis for $\mathrm{Re}\,\nu > -u$ by construction. It is easy to show that Q is analytic across the real axis for $\mathrm{Re}\,\nu < -u$ by using the identity of the two terms on the right hand side of Eq. (33). It is important that the components of $\underset{\approx}{Y}$ do not have any zeros in the complex ν-plane cut along the real axis, otherwise, STEP 2 would require additional conditions at the zeros of the components of $\underset{\approx}{Y}$. This is the reason for choosing compensating factors of the form $(\nu+u)^{-\beta_1}$ and $(\nu+u)^{-\beta_2}$ for the Wiener-Hopf factorization of D_1 and D_2 respectively (see the discussion preceeding (8 a) in Sect. I).

Then, as a consequence of the requirements of STEP 2, Q is a bounded entire function and by Liouville's theorem it must be a constant. Therefore, the problematic term (31) may formally be replaced by a term which is analytic for $\mathrm{Re}\,\nu < -u$, plus a constant, C, (perhaps $C = 0$). Together, STEP 1 and STEP 2 imply that the outcoming distribution is given by

$$\psi_0(\xi) = \frac{2}{3}\,\pi^{-1/2}(1-2u\xi)\,\underset{\approx}{\hat{X}}^{-1}(\xi,u)$$

$$\left[\int_{-u}^{\infty} \underset{\approx}{\hat{Y}}^{-1}(\xi',u)\,\frac{\xi'+u}{\xi-\xi'}\,\psi_0(\xi')\,e^{-\xi'^2}\,d\xi' + C\right], \qquad (35)$$

for $\xi < -u$,

and the resolvent of $\underset{\approx}{K}$ as a function of $\frac{1}{\nu+u}$ is analytic for $\mathrm{Re}\,\nu < -u$.

In STEP 2, the requirement at $\nu = -u$ leads to upper bounds for both β_1 and β_2 because of the two component nature of Q. Similarly, the requirement at $\nu = \infty$ leads to lower bounds for both β_1 and β_2. One must be careful to include the effect of Eq. (18) and, using an explicit expression[12] for $\underset{\approx}{\hat{Y}}$, one obtains,

$$\beta_1 = -3 \qquad (36\text{ a})$$

$$3 \le \beta_2 \le 4 \qquad \text{for} \qquad u < \sqrt{5/6} \qquad (36\text{ b})$$

and

$$3 \le \beta_2 \le 5 \qquad \text{for} \qquad u < \sqrt{5/6} \qquad (36\text{ c})$$

These conditions are the preliminary results analogous to (8 a) and (8 b) for molecules with one degree of freedom, and the supersonic case is seen to be marked in a similar manner.

It now remains to determine the consequences for each possible choice of β_2 for both the subsonic and supersonic cases. Eqs. (18), (32) and (35) are instrumental in the analysis. The idea is to require i) $\psi_0(\xi)$ or $(\xi+u)\ \psi_0(\xi)$ to be integrable for $\xi < -u$ and ii) each of the two terms on the right hand side of Eq. (32) to be asymptotically compatible for $\nu \to -u$ and $\nu \to \infty$. By analogy with Ref. (5), we would expect to obtain at most two conditions of the form,

$$\int_{-u}^{\infty} \xi'^{\,i} (\xi'+u)^j \ \hat{\underset{\approx}{Y}}(\xi',u)\ \psi_0(\xi')\ e^{-\xi'^2} d\xi' = 0 \ , \tag{37}$$

for two integer pairs (i, j), ideally corresponding to only one choice of β_2 in the subsonic case while all other choices of β_2 for either case should lead to more than two conditions of the form given by Eq. (37). However, this part of the problem is far from complete.

IV. DISCUSSION

In Sec. III, we demonstrated one method by which the resolvent of $\underset{\approx}{K}$ as a function of $\frac{1}{\nu+u}$ can be explicitly put in form which is analytic for $\mathrm{Re}\ \nu < -u$. Using the ν variable it is possible to define a contour Γ as in Fig. 1 of Ref. (5) with perhaps the addition of a finite number of imbedded singularities, so that Eq. (17) makes sense. Then, the fact that boundary condition (14 b) is satisfied follows as a direct consequence of Eq. (16) and the resolution of the identity formula[13] applied to Eq. (17).

There are, of course, other methods for solving an equation of the form (13 a) (e.g. the generalized eigenfunction method of Case[14]), and from studies of transport equations in general[15] and Vlasov plasma transport in particular[16], several methods have been shown to be equivalent. However, the particular form of the general solution to Eq. (13 a) is not crucial to the result we seek in this paper. We are not proving the existence of a unique solution to Eq. (13 a) but rather, we are investigating the consequences of requiring Eq. (13 a) to have a unique solution by using well-known properties from the theory of semigroups.

In order to see how the critical value of the speed ration ($u_{crit} = \sqrt{5/6} \leftrightarrow$ Mach number unity for molecules with three degrees of freedom) arises in our two component model of the linearized Boltzmann equation, it is necessary to consider the details of the factorization of $\underset{\approx}{\Omega}^{12}$. We have merely outlined some of these details in Sec. II.

The result sought in this paper has long been anticipated on the basis of certain physical and mathematical considerations of hydrodynamic equations[17]. An important aspect of this study is an understanding of precisely where and how the solution to the linearized Boltzmann equation breaks down for the case of supersonic flow.

AKNOWLEDGMENTS

The author wishes to thank Prof. Carlo Cercignani for suggesting this generalization of our earlier joint work and for several helpful discussions. The author is also grateful to his colleagues Dr. F. Gleisberg and Mr. Hans-Jörg Breymayer for assistance in numerical matters related to this work.

REFERENCES

1. Tor Ytrehus, article in this volume.
2. Wang Chang, C. S. and G. E. Uhlenbeck, Dept. of Engr. Research report, U. of Mich., 1952.
3. C. Cercignani, Theory and Application of the Boltzmann Equation, Scottish Academic Press, Edinburgh, and Elsevier, New York (1975).
4. Gross, E. P. and Jackson, E. A., Phys. Fluids 2, 4, 432, (1959).
5. M. D. Arthur and C. Cercignani, J. Appl. Math. and Phys. (ZAMP) Vol. 31, 634, (1980).
6. C. Cercignani, Elementary Solutions of Linearized Kinetic Models and Boundary Value Problems in the Kinetic Theory of Gases, Div. of Appl. Math. and Engr. report, Brown University, Providence, Rhode Island (1965).
7. E. W. Larsen and G. Habetler, Commun. Pure Appl. Math. 26, 525 (1973).
8. E. W. Larsen,Commun. Pure Appl. Math. 28, 729 (1975).
9. E. W. Larsen, S. Sancaktar and P. F. Zweifel, J. Math. Phys. 16, 117 (1976).
10. H. Grad, Handbuch der Physik, 12, sec. 19, (1958).
11. H. Hejtmanek, article in this volume.

12. M. D. Arthur and C. Cercignani, On the Riemann-Hilbert Problem for $\underset{\approx}{\Omega}$, in preparation.

13. Dunford, N. and Schwartz, J. T., Linear Operators, Wiley, New York (1964).

14. K. M. Case, Ann. Phys. 9, 1 (1960).

15. P. F. Zweifel, R. L. Bowden and W. Greenberg, J. Math. Phys. 5, 219 (1979).

16. M. D. Arthur, Ph. D. dissertation, available from University Microfilms, Inc., Ann Arbor, Mich. (1979).

17. C. Cercignani, Proc. Oberwolfach Conference on Math. Problems in Kinetic Theory, May 1979 (1980).

The Space-homogeneous Boltzmann Equation for Molecular Forces of Infinite Range.

Tony Elmroth

1. Introduction

Recently Arkeryd proved the existence of weak solutions to the non-linear space-homogeneous Boltzmann equation for molecular encounters with infinite range, inverse k^{th}-power forces, $k>3$, [A1]. His proof is based on the existence of solutions in the cut-off case and a weak L^1-compactness argument. A weak form is evidently enough to draw conclusions about the moments of the solutions. Arkeryd showed that for inverse k^{th}-power molecules, $k=5$, his molecular densities have moments that satisfy Ikenberry-Truesdell's equation [T] for the moments. Thus these densities converges in a weak sense to the Maxwell distribution. For $k \neq 5$ nothing similar has been proved. We have studied the case of inverse k^{th}-power molecules, $k \geq 5$. For these forces we have proved that Arkeryd's solutions have globally bounded higher moments. In this seminar we shall present the main ideas of Arkeryd's existence proof as well as our proof of the global boundedness of the higher moments.

The non-linear space-homogeneous Boltzmann equation, in a strong form is

$$(1) \qquad \begin{aligned} D_t F(t,\mathbf{v}) &= \mathbb{C}F(t,\mathbf{v}) \ , \quad t > 0 \\ F(0,\mathbf{v}) &= F_0(\mathbf{v}) \end{aligned}$$

Here the collisions operator is

$$\mathbb{C}F(\mathbf{v}) = \mathbb{C}^R F(\mathbf{v}) = \int_{\mathbb{R}^3}\int_0^R\int_0^{2\pi} w(F(\mathbf{v}')F(\mathbf{v}'_*)-F(\mathbf{v})F(\mathbf{v}_*))\,r\,dr\,d\zeta\,d\mathbf{v}_* \ ,$$

with cut-off radius R in the impact plane and the physical weight function $w = |\mathbf{v}_* - \mathbf{v}|$. A collision takes place if the impact is within the cross section $[(r,\zeta) \ ; \ 0 \leq r \leq R, \ 0 \leq \zeta \leq 2\pi]$. Then the resulting velocities $\mathbf{v}'$, $\mathbf{v}'_*$ are here determined, from the inital velocities $\mathbf{v}$, $\mathbf{v}_*$, by an inverse k^{th}-power law.

For forces of infinite range we will use a weak form of the Boltzmann equation

$$(3)\qquad \int_{\mathbb{R}^3} F(t,\mathbf{v})g(t,\mathbf{v})d\mathbf{v} = \int_{\mathbb{R}^3} F_o(\mathbf{v})g(0,\mathbf{v})d\mathbf{v} + \int_0^t \int_{\mathbb{R}^3} F(t',\mathbf{v})D_t g(t',\mathbf{v})d\mathbf{v}dt' + \int_0^t \int_{\mathbb{R}^3} ¢F(t',\mathbf{v})g(t',\mathbf{v})d\mathbf{v}dt' , \quad g \in C^{1,\infty} .$$

Here $C^{1,\infty} = [g \in C^1([0,\infty)\times\mathbb{R}^3);\ |g|_1 = \sup|g| + \sup|D_t g| + \sup|\mathrm{grad}_v g| < \infty]$. (3) is formally obtained by multiplying (1) with testfunctions $g \in C^{1,\infty}$, integrating in t and $\mathbf{v}$ and carrying out a partial integration in t .

We define a solution of (3) to be any function $F\colon R_+ \to L^1_+$, $t \to F(t,\mathbf{v})$, such that (3) is valid for all $g \in C^{1,\infty}$. Here $R_+ = [t \in \mathbb{R} ; t \geq 0]$ and L^1_+ is the positive cone in L^1, $L^1_+ = [f \in L^1 ; f \geq 0]$. The form of the collisions integral that will be used in (3) is

$$(4)\qquad \int_{\mathbb{R}^3} ¢F(\mathbf{v})g(\mathbf{v})d\mathbf{v} = \int_{\mathbb{R}^3}\int_{\mathbb{R}^3}\int_0^\infty\int_0^{2\pi} (g(\mathbf{v}')-g(\mathbf{v}))wF(\mathbf{v})F(\mathbf{v}_*)rdrd\zeta d\mathbf{v}d\mathbf{v}_* .$$

For molecules with cut-off (4) can be obtained from (2) by a change of variables. For molecules without cut-off we take (4) as our definition of the collisions integral.

Recall the following result about solutions in the cut-off case.

<u>Proposition 1.</u> [A2] For $¢ = ¢^R$, $R < \infty$, the Boltzmann equation (1) has a unique solution $F\colon R_+ \to L^1_+$, $t \to F(t,\mathbf{v})$, for every $F_o \in L^1_+$ such that $(1+v^4)F_o \in L^1_+$. The solution has the following properties:

1. F conserves mass, momentum and energy

$$\int_{\mathbb{R}^3} g(\mathbf{v})F(t,\mathbf{v})d\mathbf{v} = \int_{\mathbb{R}^3} g(\mathbf{v})F_o(\mathbf{v})d\mathbf{v} , \quad t>0$$

for $g(\mathbf{v}) = 1, \mathbf{v}, v^2$.

2. If $F_o \log F_o \in L^1$ then $F(t)\log F(t) \in L^1$ for all $t>0$ and

$$HF(t) = \int_{\mathbb{R}^3} F(t,\mathbf{v}) \log F(t,\mathbf{v})\, d\mathbf{v}$$

is non-increasing as a function of t .

3. If $(1+v^s)F_o \in L^1$ then $(1+v^s)F(t) \in L^1$, for all $t > 0$.

Letting the cut-off go to infinity, in Proposition 1, and taking the limit, we can obtain the following theorem.

Theorem 2.[A1] For inverse k^{th}-power molecules, with $k > 3$ and initial distribution $F_o \in L^1_+$, such that $(1+v^4)F_o \in L^1_+$, $F_o \log F_o \in L^1$, there exists a weak solution $F: R_+ \to L^1_+$ to the Boltzmann equation (3). This solution conserves mass, momentum and energy.

Remark. E.g. using bounded weight function $w_m = \max(w,m)$ and letting m go to infinity together with R , it is possible to obtain the existence with only $(1+v^2)F_o \in L^1_+$, $F_o \log F_o \in L^1$,[A1].

From global bounds of higher moments in the cut-off case, which are independent of the cut-off radius, we obtain the following theorem about the corresponding moments for k^{th}-power forces of infinite range.

Theorem 3. [E] If $k \geq 5$ and $(1+v^{s+\varepsilon})F_o \in L^1_+$ for some $\varepsilon > 0$, then the solutions F of Theorem 2 have globally bounded moments of order s , i.e. there exists a constant $C_{F_o,s}$ only depending on F_o and s , such that

$$\int_{\mathbb{R}^3} v^s F(t,v)dv \leq C_{F_o,s} \text{ , for all } t > 0 \text{ .}$$

Theorem 4.[A1,E] For $k > 3$ the solutions F of Theorem 2 satisfy: $(1+v^s)F(t) \in L^1_+$ for all $t > 0$, if $(1+v^s)F_o \in L^1_+$.

We will give the main ideas of the proofs of Theorem 2, and Theorem 3 , when $s = 2n$ and $\varepsilon \geq (k-5)/(k-1)$.

2. Proof of Theorem 2

For the proof of Theorem 2 the following lemma is essential.

Lemma 5. Let $(F_n)_1^\infty$ be a sequence of L^1_+-functions such that for some $s > 0$,

$$\int_{\mathbb{R}^3} F_n(v)(1+v^2)^s dv < C_s , \quad n = 0, 1, 2, \ldots ,$$

and such that

$$\int_{\mathbb{R}^3} F_n(v) \log F_n(v) dv < C, \; n = 0, 1, 2, \ldots .$$

Then $(F_n)_1^\infty$ contains a sub-sequence $(F_{n_j})_{j=1}^\infty$ converging weakly to a function $F \in L_+^1$. Moreover; If $(1+v^2)^{-s'} g(v) \in L^\infty$ for some s', $0 \le s' < s$, then

$$\lim_{j \to \infty} \int_{\mathbb{R}^3} F_{n_j}(v) g(v) dv = \int_{\mathbb{R}^3} F(v) g(v) dv .$$

For a proof of the Lemma see e.g. [A2].

<u>Proof of Theorem 2</u>. The solutions of (1) from Propostion 1 satisfies the conditions in Lemma 5 with $s = 1$ for all $t > 0$. Let $(F_R(t,\cdot))_{R=1}^\infty$ be a sequence of solutions to (1) with cut-off radius R. By Lemma 5 we can choose a subsequence $F_{R_j}(t,\cdot)$ which converges in weak L^1-sense to a non-negative L^1-function $F(t,\cdot)$ for all rational t. Assuming the equicontinuity of

$$(5) \qquad \Big(\int_{\mathbb{R}^3} F_{R_j}(t,v) g(t,v) dv \Big)_{j=1}^\infty , \quad g \in C^{1,\infty} ,$$

the convergence for all t can be proved as follows. Supposing the contrary, that there is a non-rational t_o such that $F_{R_j}(t_o)$ does not converge in weak L^1-sense, then for some $g \in C^{1,\infty}$ the sequence (5) will not converge. This would by the t-equicontinuity contradict the weak convergence for rational t. Thus

$$F_{R_j}(t,\cdot) \to F(t,\cdot) , \text{ for all } t ,$$

in weak L^1-sense.

Now the equicontinuity of (5) follows from

$$(6) \qquad \Big| \int_{\mathbb{R}^3} F_{R_j}(t_2,v) g(t_2,v) dv - \int_{\mathbb{R}^3} F_{R_j}(t_1,v) g(t_1,v) dv \Big| \le$$

$$\le \Big| \int_{t_1}^{t_2} \int_{\mathbb{R}^3} F_{R_j}(t',v) D_t g(t',v) dt' dv + \int_{t_1}^{t_2} \int_{\mathbb{R}^3} ¢^{R_j} F_{R_j}(t',v) g(t',v) dt' dv \Big| \le$$

$$\leq |t_2 - t_1|\ |g|_1 \int_{\mathbb{R}^3} F_o(\mathbf{v})d\mathbf{v} + |t_2 - t_1|\ |g|_1 C\Big[\int_{\mathbb{R}^3} (1+v^2)F_o(v)d\mathbf{v} \Big]^2,$$

where we have used (3) and the following estimate of the collisions integral.

In angular coordinates (see e.g. [TM] p. 97) a scattering factor S can be defined as

$$\sin\theta \quad S(\theta,w) = wr(\theta,w)|D\ r(\theta,w)| \quad .$$

We will use $\mathbf{v}' = \mathbf{v}'(\theta)$ and

(7) $$|g(\mathbf{v}')-g(\mathbf{v})| \leq \sup|\mathrm{grad}_{\mathbf{v}'}g(\mathbf{v}')|w(\pi/2-\theta) \ .$$

We can estimate the collisions integral (4) as

(8) $$\Big| \int \mathbb{C}F(t',\mathbf{v})g(t,\mathbf{v}) \Big| \leq$$

$$\leq \int_{\mathbb{R}^3}\int_{\mathbb{R}^3}\int_0^{\pi/2}\int_0^{2\pi} F(t,\mathbf{v})F(t,\mathbf{v}_*)(\pi/2-\theta)w\ \sin\theta\ S(\theta,w)d\theta d\zeta d\mathbf{v}d\mathbf{v}_* \cdot|g|_1 \quad .$$

For inverse k^{th}-power molecules, $k > 3$, we have

(9) $$\int_0^{\pi/2} (\pi/2-\theta)w\ \sin\theta\ S(\theta,w) \leq \quad C\cdot w^{(2k-6)/k-1)} \leq C\cdot(1+v)^2(1+v_*)^2 \ .$$

This gives for the collisions integral

$$\Big| \int_{\mathbb{R}^3} \mathbb{C}F(t,\mathbf{v})g(t,\mathbf{v})d\mathbf{v} \Big| \leq C\cdot\int_{\mathbb{R}^3}\int_{\mathbb{R}^3} (1+v)^2(1+v_*)^2F(t,\mathbf{v})F(t,\mathbf{v}_*)d\mathbf{v}d\mathbf{v}_*|g|_1 \leq$$

$$\leq C\cdot\Big[\int_{\mathbb{R}^3}\int_{\mathbb{R}^3} (1+v^2)F_o(\mathbf{v})d\mathbf{v} \Big]^2\cdot|g|_1 \ ,$$

which is the estimate used in (6) . For the proof of (7) and (9) see e.g. [TM] .

By (7) and (8) we can choose F_{R_j} , so that

$$\int_0^t \int_{\mathbb{R}^3} \mathbb{C}^{R_j}F_{R_j}(t',\mathbf{v})g(t',\mathbf{v})dt'd\mathbf{v} \to \int_0^t \int_{\mathbb{R}^3} \mathbb{C}F(t',\mathbf{v})g(t',\mathbf{v})dt'd\mathbf{v} \ , j\to\infty.$$

Taking the limit in

$$\int_{\mathbb{R}^3} F_{R_j}(t',\mathbf{v})g(t',\mathbf{v})dt'd\mathbf{v} = \int_{\mathbb{R}^3} F_o(\mathbf{v})g(0,\mathbf{v})d\mathbf{v} +$$

$$+ \int_0^t \int_{\mathbb{R}^3} F_{R_j}(t',v) d_t g(t',v) dt' dv + \int_0^t \int_{\mathbb{R}^3} ¢^{R_j} F_{R_j}(t',v) g(t',v) dt dv \ ,$$

$j \to \infty$, we conclude that F is a solution of (3), i.e. of the Boltzmann equation in the case of infinite range forces.

For $s' < 1$

$$\lim_{j \to \infty} \int_{\mathbb{R}^3} (1+v^2)^{s'} F_{R_j}(t,v) dv = \int_{\mathbb{R}^3} (1+v^2)^{s'} F(t,v) dv \ ,$$

and the 0th and 1st moments are conserved. For the energy we have

$$\int_{\mathbb{R}^3} v^2 F_{R_j}(t,v) dv = \int_{\mathbb{R}^3} v^2 F_o(v) dv \ , \ t > 0 \ , \ j = 1, 2, 3, \ldots$$

which yields

$$\int_{\mathbb{R}^3} v^2 F(t,v) dv \leqslant \int_{\mathbb{R}^3} v^2 F_o(v) dv \ .$$

To prove conservation of the energy Theorem 4 can be used,(see [A1,E]).

3. Proof of Theorem 3, for $s = 2n$ and $\varepsilon \geq (k-5)/(k-1)$.

Multiplying the Boltzmann equation (1) with v^{2n} and integrating we obtain the following equation for the moment of order 2n, $M_F(2n)$.

$$D_t M_{F(t)}(2n) = \int_{\mathbb{R}^3} ¢^R F(t,v) v^{2n} dv \ .$$

In the next section we estimate the collisions integral,

$$\int_{\mathbb{R}^3} ¢^R F_R v^{2n} dv \leq P(\text{moments of order} \leq 2n-2) - C \cdot M_{F_R}(2n) \ , \tag{10}$$

where P is a pollynomial and C a positive constant, both independent of R. Using this estimate and assuming all moments $M_{F_R}(\cdot)$ of order less than or equal to 2n-2 are globally bounded, we obtain

$$D_t M_{F_R}(2n) \leq K_{2n-2} - C \cdot M_{F_R}(2n),$$

$$M_{F_R}(2n) \leq M_{F_o}(2n) e^{-Ct} + (K_{2n-2}/C)(1-e^{-Ct}) \ ,$$

thus $M_{F_R}(2n) \leq C_{2n}$ for all $t > 0$, where C_{2n} only depends on moments of F_o and is independent of R. F_R has constant energy. By induction all higher moments that exists, under the restrictions in Theorem 3, are globally bounded.

For a weak solution F constructed as the weak L^1 limit of the sequence (F_{R_j}) we obtain

$$\int_{\mathbb{R}^3} F(t,v)v^{2n}dv \leq C_{2n} \ , \ t > 0 \ ,$$

Thus the solution F has globally bounded moments.

4. Estimate of the collisions integral.

We start with some geometric considerations of the collisions to obtain estimates of

$$v'^{2n} + v_*'^{2n} - v^{2n} - v_*^{2n} \ . \tag{11}$$

The velocities before collision are v, v_*. For each relative velocity $w = v_* - v$ we fix an ON-system w/w, $\xi(w)$, $\eta(w)$ with $\xi(w)$ and $\eta(w)$ in the plane perpendicular to w. The velocities after collisions are given by

$$\begin{aligned} v' &= v + (w\cdot a)a \ , \\ v_*' &= v - (w\cdot a)a \ , \end{aligned} \tag{12}$$

where a is a unit vector in the direction of change of momentum

$$a = \frac{v' - v}{|v' - v|} .$$

Define θ and ζ through

$$a = a(\theta,\zeta,w) = \cos\theta \ w/w - \sin\theta \ \cos\zeta \ \xi(w) - \sin\theta \ \sin\zeta \ \eta(w). \tag{13}$$

For each $\theta \in [0,\pi/2]$, $\zeta \in [0,2\pi)$ we get a posible solution of the encounter problem. We now introduce $a^{\perp} = \frac{v_*' - v}{|v_*' - v|}$, a unit vector in the same plane as a and w and orthogonal to a,

$$a^{\perp} = \sin\theta \ w/w + \cos\theta \ \cos\zeta \ \xi(w) + \cos\theta \ \sin\zeta \ \eta(w) \ .$$

a, $a^{\perp}$, and $a \times a^{\perp}$ will be an ON-system

$$a \times a^{\perp} = -\sin\theta \ \xi(w) + \cos\zeta \ \eta(w) \ .$$

Using this ON-system and (8) to express v, v_*, v', and v_*', we obtain

$$\begin{aligned} v &= {}_1v \ a + {}_2v \ a^{\perp} + {}_3v \ a \times a^{\perp} \ , \\ v_* &= {}_1v_* a + {}_2v_* a^{\perp} + {}_3v \ a \times a^{\perp} \ , \end{aligned}$$

$$\mathbf{v}' = {}_1v_* \mathbf{a} + {}_2v\, \mathbf{a}^{\perp} + {}_3v\, \mathbf{a}\times\mathbf{a}^{\perp} ,$$
$$\mathbf{v}'_* = {}_1v\, \mathbf{a} + {}_2v_* \mathbf{a}^{\perp} + {}_3v\, \mathbf{a}\times\mathbf{a}^{\perp} .$$

For (11) we obtain

$$v'^{2n} + v'^{2n}_* - v^{2n} - v_*^{2n} = ({}_1v_*^2 + {}_2v^2 + {}_3v^2)^n +$$
$$+ ({}_1v^2 + {}_2v_*^2 + {}_3v^2)^n - ({}_1v^2 + {}_2v^2 + {}_3v^2)^n -$$
$$- ({}_1v_*^2 + {}_2v_*^2 + {}_3v^2)^n =$$
$$= \sum_{\nu=1}^{n-1} \binom{n}{\nu} {}_1v_*^{2\nu}\, {}_2v^{2(n-\nu)} + \sum_{\nu=1}^{n-1} \binom{n}{\nu} {}_1v^{2\nu}\, {}_2v_*^{2(n-\nu)} -$$
$$- \sum_{\nu=1}^{n-1} \binom{n}{\nu} {}_1v^{2\nu}\, {}_2v^{2(n-\nu)} - \sum_{\nu=1}^{n-1} \binom{n}{\nu} {}_1v_*^{2\nu}\, {}_2v_*^{2(n-\nu)} ,$$

where the binomial expansion has been used. Estimating the two first terms from above and the two last from below we can obtain

$$(14) \qquad v'^{2n} + v'^{2n}_* - v^{2n} - v_*^{2n} \leq n! \sum_{\nu=1}^{n-1} v^{2\nu} v_*^{2(n-\nu)} - C_\varepsilon v^{2n} .$$

This holds under the following restrictions on θ and ζ,

$$\theta \in \Theta = [\theta : \varepsilon \leq \theta \leq \pi/2 - \varepsilon] ,$$
$$\zeta \in Z = [\zeta : |\sin(\zeta - \zeta_{\mathbf{v}})| > \sin\varepsilon] , \ 0 < \varepsilon < 10^{-1} .$$

For the rest of the proof we fix such an ε .

To estimate the collisions integral (10) we use the scattering factor S and split the integration in θ into two parts

$$\int_{R^3} \mathbb{C}^R F(\mathbf{v}) v^{2n} d\mathbf{v} =$$
$$= \int_{R^3}\int_{R^3}\int_0^m\int_0^{2\pi} (v'^{2n} - v^{2n}) F(\mathbf{v}) F(\mathbf{v}_*) w S(\theta,w) \sin\theta \; d\theta d\zeta d\mathbf{v} d\mathbf{v}_* +$$
$$+ \int_{R^3}\int_{R^3}\int_m^A\int_0^{2\pi} (v'^{2n} - v^{2n}) F(\mathbf{v}) F(\mathbf{v}_*) w S(\theta,w) \sin\theta \; d\theta d\zeta d\mathbf{v} d\mathbf{v}_* =$$
$$= I_1(\delta) + I_2(\delta) ,$$

where $A = A_R(w)$ is the cut -off angle for cut-off radius R and relative velocity w , and $m = \min(\delta/2-\pi)$. In the first of these integrals $I_1(\delta)$ we make a change of variables and use $w\sin\theta \; S(\theta,w) = Kw^{(k-5)/(k-1)} \cdot z(\theta) z'(\theta)$ and inequality (14).

$$I_1(\delta) = \frac{1}{2}\int_{R^3}\int_{R^3}\int_0^{m}\int_0^{2\pi} (v'^{2n}+v_*'^{2n}-v_*^{2n}-v^{2n})F(v)F(v_*)w^{(k-5)/(k-1)}Kzz'd\theta d\zeta dv dv_* \le$$

$$\le n!\int_{R^3}\int_{R^3}\int_0^{\pi/2-\delta}\int_0^{2\pi}\sum_{\nu=1}^{n-1} v^{2\nu}v_*^{2(n-\nu)}w^{(k-5)/(k-1)}F(v)F(v_*)Kz(\theta)z'(\theta)d\theta d\zeta dv dv_* -$$

$$- C_\varepsilon/2\iint_{w>w_o}\int_\varepsilon^{\pi/2-\varepsilon}\int_Z v^{2n}w^{(k-5)/(k-1)}F(v)F(v_*)Kz(\theta)z'(\theta)d\theta d\theta dv dv_* ,$$

for $R > R_o$, where R_o and w_o are chosen so that $A_{R_o}(w_o) > \pi/2-\varepsilon$. The last integral can be estimated from below by $C\cdot M_F(2n+(k-5)/(k-1))-C_o$ where C and C_o will depend on the moments $M_F(0)$ and $M_F(2)$ but will be independent of δ. For $I_1(\delta)$ we now obtain

$$I_1(\delta) \le K_1(\delta)\sum_{\nu=1}^{n-1} M_F(2\nu+(k-5)/(k-1))M_F(2(n-\nu)) -$$
$$- C\cdot M_F(2n+(k-5)/(k-1)) + C_o .$$

The integral $I_2(\delta)$ is estimated in the same way as (8) in the proof of Theorem 2.

$$I_2(\delta) \le \int_{R^3}\int_{R^3} v^{2n-1}ww^{(k-5)/(k-1)}F(v)F(v_*)dv dv_* \cdot$$
$$\int_{\pi/2-\delta}^{\pi/2}\int_0^{2\pi}(\pi/2-\theta)z(\theta)z'(\theta)d\theta d\zeta \le K_2(\delta)M_F(2n+(k-5)/(k-1)) ,$$

where $K_2(\delta) \to 0$ when $\delta \to 0$.

Using these estimates on the solutions F_R of (1), which have constant 0th and 2nd moments, we obtain with δ_o such that $K_2(\delta_o) < C/2$,

(15)
$$\int_{R^3} \phi^R F_R(t,v)dv \le K(\delta_o)\sum_{\nu=1}^{n-2} M_{F_R(t)}(2\nu+(k-5)/(k-1))M_{F_R(t)}(2(n-\nu)) +$$
$$+ K(\delta_o)M_{F_R(t)}(2(n-1)+(k-5)/(k-1))M_{F_o}(2) -$$
$$- C/2\ M_{F_R(t)}(2n+(k-5)/(k-1)) + C_o ,\quad R > R_o .$$

Now there exists a constant C' such that, $K_a v^a - K_b v^b < C'$, $b>a$, which yields

$$K(\delta_o)M_{F_R(t)}(2(n-1)+(k-5)/(k-1))M_{F_o}(2) -$$
$$- C/4\ M_{F_R(t)}(2n+(k-5)/(k-1)) \le C'M_{F_o}(0) .$$

This together with (15) gives (10).

We have now completed the sketch of the proof of Theorem 3 in the case $s = 2n$ and $\varepsilon \geq (k-5)/(k-1)$. For the full proof see [E].

References.

[A1]. L. Arkeryd, Intermolecular Forces of Infinite Range and the Boltzmann Equation. Arch. for Rational Mech. and Anal., 77(1981), 11-21

[A2]. L. Arkeryd, On the Boltzmann Equation. Arch. for Rational Mech. and Anal., 45 (1972), 1-34.

[T] . C. Truesdell, On the pressures and the flux of energy in a gas according to Maxwell's kinetic theory, II , Jour. of Rational Mech. and Anal., 5 (1956), 55-128.

[TM]. C.Truesdell and R.G. Muncaster, Fundamentals of Maxwell's Kinetic Theory of a Simple Monatomic Gas , Academic Press (1980).

[E] . T. Elmroth, Global Boundeness of Moments of Solutions of the Boltzmann Equation for Infinite Range Forces , Tech.Report 1981 Dep. of Math. Chalmers University of Technology,Gothenburg.

THE CAUCHY PROBLEM FOR THE BOLTZMANN EQUATION.
A SURVEY OF RECENT RESULTS

Andrzej Palczewski

Institute of Mechanics
University of Warsaw
00-901 Warsaw, PKiN IX p.

Abstract. The initial value problem for the Boltzmann equation is discussed. A survey of last decade results is presented.

INTRODUCTION

The aim of these notes is the presentation of recently proved theorems on the existence and uniqueness of the solution of the Boltzmann equation. I restrict myself to the Cauchy problem only. However from physical point of view boundary value problems are more important, imposing of boundaries makes a mathematical presentation of the problem less clear and technically complicated. From the point of view of existence theorems boundary problems have no real advantage since what can be proved for problems with boundaries can be proved for problems without boundaries as well.

The Cauchy problem for the Boltzmann equation is:

(1) $$f_t + v \cdot \nabla_x f = J(f,f), \quad f(t=0) = f_o ,$$

where

$$f = f(x,v,t), \quad x \in R^3, \quad v \in R^3, \quad t \in R^+$$

and J is the Boltzmann collision operator, which has the form:

(2) $$J(f,f) = Q(f,f) - P(f,f)$$

and

$$Q(f,f) = \int_{R^3}\int_{S^2} k(v,v_1;u) f(v') f(v_1') dv_1 du ,$$

$$P(f,f) = \int_{R^3}\int_{S^2} k(v,v_1;u) f(v) f(v_1) dv_1 du ,$$

where

$$v' = v + u(u\cdot V) \ ,$$
$$v_1' = v_1 - u(u\cdot V) \ ,$$
$$V = v - v_1$$
$$u \in S^2 = \{u : |u| = 1\} \ .$$

In the presentation following topics will be taken into account.

1. Model of particle interactions.

Form of the collision kernel $k(v,v_1;u)$ in (2) is strongly dependent on the model of the particle interactions. It is assumed that collision kernel can be estimated as follows [2,3,4]:

$$(3) \qquad k(v,v_1;u) \le C\ (1 + |v|^{\lambda_1} + |v_1|^{\lambda_2}), \qquad -3 < \lambda_1, \quad \lambda_2 < 2.$$

This model includes hard spheres collisions and interactions by inverse power law potentials with Grad's angular cut-off [1]. The particular case of Maxwell molecules $\lambda_1 = \lambda_2 = 0$ can be distinguished in this model. In this case collision kernel is bounded and all considerations are much simpler than in other cases.

2. Space of functions.

The usual way in which equation (1) is treated is considering it as an ordinary differential equation in a Banach space of functions of the variables (x,v). In what follows we shall consider as such spaces $L_\infty(R^6)$, $L_1(R^6)$ and as an intermediate space $L_{1,\infty} = L_\infty(x) \times L_1(v)$, i.e. $L_{1,\infty} = L_\infty(R^3, L_1(R^3))$. Let us observe that from the physical point of view the most natural is the space $L_1(R^6)$ as f is a particle density and should be an integrable function. /In fact physically natural is integrability in the velocity space and local integrability in the configuration space./ The space $L_{1,\infty}$ can be interpreted as a space of finite mass densities and only the space $L_\infty(R^6)$ has no simple physical interpretation. However the space $L_\infty(R^6)$ will be considered with a weight ρ, which assures that it is a subspace of $L_{1,\infty}$ / we shall denote it by $L_\infty(R^6,\rho)$.

3. Smoothness of solution.

Since we consider equation (1) in an abstract form as an ordinary differential equation in a Banach space solution of this equation can be either a strong solution or a mild solution of the problem. In the first case solution is a strongly differentiable function with respect to t and in the second it is only a continuous function.

RESULTS

The first result I would mention is following theorem of Maslova and Tchubenko [3]:

Theorem 1. Let us assume that

1/ the collision kernel is such that estimation (3) holds with $-1 < \lambda_1, \lambda_2 \leq 1$,

2/ equation (1) is considered in the space $L_\infty(R^6,\rho)$ with the weight $\rho = (1+|v|^2)^{s/2}$, $s > 6$.

Then for the initial data $f_o \in L_\infty(R^6,\rho)$ exists $T > 0$ such that for $t \in [0,T]$ equation (1) has unique mild solution $f(t) \in L_\infty(R^6,\rho)$.

Theorem 1 can be proved using the fixed point argument. The main step is the estimation of the collision integral given by following lemma [3]:

Lemma 1. Under the assumption of theorem 1 we have:

$$\|Q(f,f)[\gamma + P(f,f)]^{-1}\|_\infty \leq \|f\|_\infty (k_1 + k_2\|f\|_\infty), \qquad \gamma > 0,$$

where $\|\cdot\|_\infty$ is the norm in $L_\infty(R^6,\rho)$.

Extension of theorem 1 on a wider class of interaction models and in a slightly different space was given by Kaniel and Shinbrot [2]:

Theorem 2. Let us assume that:

1/ the collision kernel is such that estimation (3) holds with $-3 < \lambda_1, \lambda_2 < 2$,

2/ equation (1) is considered in the space $L_\infty(R^6,\rho)$ with the weight $\rho = \exp b|v|^2$, $b > 0$.

Then for the initial data $0 \leq f_o \in L_\infty(R^6,\rho)$ exists $T > 0$ such that for $t \in [0,T]$ equation (1) has unique mild solution $0 \leq f(t) \in L_\infty(R^6,\rho)$.

Technics used in the proof of theorem 2 are completely different from that used for theorem 1. Namely with the nonlinear equation (1) following pair of successive approximations can be associated:

$$(4) \qquad \begin{aligned} u^n_t + v\cdot\nabla_x u^n + P(u^n, l^{n-1}) &= Q(u^{n-1}, u^{n-1}), \\ l^n_t + v\cdot\nabla_x l^n + P(l^n, u^{n-1}) &= Q(l^{n-1}, l^{n-1}). \end{aligned}$$

Lemma 2 [2]. Under the assumptions of theorem 2 sequences defined by (4) exist in $L_\infty(R^6,\rho)$ and if

$$0 \le l^o \le l^1 \le u^1 \le u^o$$

then for all n holds

$$0 \le l^o \le l^1 \le \ldots \le l^n \le \ldots \le u^n \le \ldots \le u^1 \le u^o .$$

Consequently, $\{u^n\}$ and $\{l^n\}$ converge in $L_\infty(R^6,\rho)$.

We proceed now to the case of the existence of the strong solutions in the space L_1. The price we shall pay for this generalization is a restricted model of interactions.

Theorem 3.[4]. Let us assume that:

1/ the collision kernel is such that estimation (3) holds with $\lambda_1 = \lambda_2 = 0$ /Maxwell molecules/ ,

2/ equation (1) is considered in the space $L_1(R^6)$.

Then for the initial data $0 \le f_o \in L_1(R^6) \cap L_{1,\infty}(R^6)$ exists $T > 0$ such that for $t \in [0,T]$ equation (1) has unique strong solution $f(t)$ in L_1 and $0 \le f(t) \in L_1(R^6) \cap L_{1,\infty}(R^6)$.

Proof of this theorem is based on the following estimation of the collision kernel [4]:

Lemma 3. Under the assumptions of theorem 3 we have:

$$\|J(f,f)\|_{1,\infty} \le C[\|f\|_{1,\infty}]^2 ,$$

where $\|\cdot\|_{1,\infty}$ is the norm in $L_{1,\infty}$.

CONCLUSIONS

Presented theorems on the existence and uniqueness of solution seem to be nonsatisfactory. None of these theorems solves the problem completely. The theorem which says that unique strong solution of the Boltzmann equation in the space L_1 for non-Maxwellian molecules exists is still needed. It is the author's opinion that the easiest way is a generalization of theorem 3 on wider class of intermolecular potentials.

We omit in this discussion the most serious problem in the existence theory for the Boltzmann equation. Namely the problem of global existence of solution i.e. existence of solution for all times $t \ge 0$. The author does not believe that presented theorems show a proper way to the solution of this problem. The main step on this way should be a better estimation of the collision integral and it is hardly believable that estimations better than given in Lemmas 1 and 3 are available. As a byproduct of theorem 3 we have corollary which says that prolongation of the solution on a greater time interval can fail only as a result of unlimited increase of the mass density / $\|\cdot\|_{1,\infty}$ - norm measures a mass density of solution! / . It shows that lack of the global solution can be a property of the model itself.

REFERENCES

[1] C. Cercignani, Theory and Application of the Boltzmann Equation, Scottish Academic Press, 1975.

[2] S. Kaniel, M. Shinbrot, The Boltzmann equation. Uniqueness and local existence, Comm. Math. Phys. 58 (1978), 65-84 .

[3] N.B. Maslova, R.P. Tchubenko, Solutions of the nonstationary Boltzmann equation (in Russian), Vestnik Len. Univ. 1973 nr. 19, 100-105.

[4] A. Palczewski, Local existence theorem for the Boltzmann equation in L_1, Arch. of Mech., 33 (1981), 973-981.

BOLTZMANN HIERARCHY AND BOLTZMANN EQUATION

Herbert SPOHN

1. Introduction.

When deriving for a system of particles at low density the Boltzmann equation from the BBGKY hierarchy, one is led rather naturally to a linear system of equations for the scaled correlation functions at zero density [1,2,3]. This set of equations has been called the Boltzmann hierarchy. The derivation of the Boltzmann equation is then completed in a second step. One observes that, if the initial data for the Boltzmann hierarchy factorize as $r_n = \prod_{j=1}^{n} r$, then so does its solution $r_n(t) = \prod_{j=1}^{n} r(t)$, $n = 1,2,\ldots$, and $r(t)$ is the solution of the non-linear Boltzmann equation at time t with initial data r . This property of the Boltzmann hierarchy is usually called "propagation of (the initial) chaos".

Having obtained the Boltzmann hierarchy poses immediately two problems. (i) Is the Boltzmann hierarchy a generalization of the Boltzmann equation ? (ii) Are there conceivably physical situations in which chaos (factorization) is not satisfied and consequently the Boltzmann hierarchy rather than the Boltzmann equation has to be used ? The answer to the first question is in a way implicitly contained in the work of O.E. Lanford on the derivation of the Boltzmann equation. Since to my experience even specialists in the field do not know the resolution of this problem, I hope it to be appropriate for a Summer School on the "Boltzmann Equation" to explain the simple answer : The solutions of the Boltzmann hierarchy are precisely statistical solutions of the Boltzmann equation. To the second question I have no convincing example. Instead I will offer in the last section some speculations based on the analogy with dissipative dynamical systems.

2. An Example.

Not to loose ourselves in large spaces it is of importance to keep a simple example in mind. I want to consider the ordinary differential equation

$$\frac{d}{dt} x(t) = x(t) - x(t)^2 . \tag{2.1}$$

To have a compact state space I will restrict myself to $x(0) = x \in \Omega = [1/2,3/2]$

and to $t \geq 0$. Then the solution to (2.1) defines the semiflow $T_t : \Omega \to \Omega$ by $x(t) = T_t x$ . . The semiflow has one globally attracting fixed point, $\lim_{t\to\infty} T_t x = 1$ for all $x \in \Omega$. The evolution of smooth functions is given by

$$f(x,t) = f(T_t x)$$

and they satisfy the (backwards) drift equation

$$\frac{\partial}{\partial t} f(x,t) = (x-x^2) \frac{\partial}{\partial x} f(x,t) . \tag{2.2}$$

Suppose we have a situation, where the initial data are not fixed (not deterministic), but rather statistical and distributed according to the probability measure μ on Ω . Signed measures form the space dual to $C(\Omega)$, the space of bounded and continuous functions on Ω . Therefore the time evolved measure at time t (the statistical distribution at time t) is defined through

$$\mu_t(f) = \mu(f \circ T_t) \tag{2.3}$$

for all $f \in C(\Omega)$, where $\mu(f) = \int \mu(dx) f(x)$. If $\mu_t(dx)$ has a smooth density, $\mu_t(dx) = \rho(x,t)dx$, then the density satisfies the (forward) drift equation dual to (2.2),

$$\frac{\partial}{\partial t} \rho(x,t) = - \frac{\partial}{\partial x} [(x-x^2)\rho(x,t)] . \tag{2.4}$$

To describe the evolution of a measure under the semiflow T_t one may instead of (2.4) also consider the differential equations for the moments of μ_t . This appears to be particularly convenient, since (2.1) has only a quadratic nonlinearity. The moments are defined by

$$r_n(t) = \mu_t(x^n) = \int \mu(dx)(T_t x)^n .$$

Since Ω is compact and since $T_t x > 0$,

$$0 \leq r_n(t) \leq z^n$$

with $z = 3/2$. By Stone-Weierstrass the polynomials are dense in $C(\Omega)$ and therefore the moments $r_n(t)$ determine μ_t uniquely. The moments satisfy the coupled set of ordinary differential equations

$$\frac{d}{dt} r_n(t) = n\, r_n(t) - n\, r_{n+1}(t) \,, \qquad (2.5)$$

$$n = 0,1,\ldots \,.$$

(2.5) satisfies the "propagation of chaos". If $r_n(0) = \lambda^n$, $n = 0,1,\ldots$, $\lambda \in \Omega$, then the solution of (2.5) is $r_n(t) = \lambda(t)^n$ with $\lambda(t) = T_t\lambda$. In fact, since $r_n(0)$ are moments, it suffices to assume that $r_2(0) = r_1(0)r_1(0)$.

The analogy to the Boltzmann hierarchy is now quite apparent. The choice of spaces is dictated by Landford's theorem as to be discussed in the following section.

state space Ω ⤳ $\{r : \Lambda \times \mathbb{R}^3 \to \mathbb{R} \,|\, r$ is measurable and $0 \le r(q,p) \le z h_\beta(p)\}$

differential equation (2.1) ⤳ Boltzmann equation with the restriction that $0 \le t \le 0.1\sqrt{\beta}/z$

moment equations (2.5) ⤳ Boltzmann hierarchy.

3. Hard Sphere Gas / Grad Limit / Random Field.

We consider a gas of hard spheres of diameter εd and unit mass, moving inside the box $\Lambda \times \mathbb{R}^3$. For notational simplicity we assume periodic boundary conditions. Since the number of particles is not necessarily fixed, the classical phase space is $\Gamma = \bigcup_{n \ge 0} (\Lambda \times \mathbb{R}^3)^n$. A point in Γ is specified by the number n of particles and their positions and momenta $(q_1,p_1,\ldots,q_n,p_n)$, $q_j \in \Lambda$, $p_j \in \mathbb{R}^3$. As shorthand we use $x_j = (q_j,p_j)$. Γ contains points where spheres overlap. These will be excluded by assigning to them initially and thereby at any time probability zero. We assume that the initial state of the system is given by a probability measure μ^ε on Γ . μ^ε is assumed to have densities, $\mu^\varepsilon(dx_1 \ldots dx_n) = f_n(x_1,\ldots,x_n) \frac{1}{n!} dx_1 \ldots dx_n$, and vanishes whenever two spheres are closer than εd . As long as spheres do not touch each other they move freely. Upon touching they collide elastically according to the laws of mechanics. This prescription defines a (Hamiltonian) flow T_t^ε on the phase space Γ . T_t^ε sends the initial phase point to the phase point at time t . The time evolved measure is given by $\mu^\varepsilon \circ T_{-t}^\varepsilon$. [In the course of time grazing and triple collisions may occur. Beyond these the dynamics remains undefined. A theorem due to Alexander [4] ensures that these exceptional situations do not occur too often. He proves that they form a set of Lebesgue measure zero, cf. also [5].]

We want to investigate the system of hard spheres at low (volume) density. The appropriate scaling is the Grad limit

diameter = εd

number of particles $\sim \varepsilon^{-2}$

with $\varepsilon \to 0$. The Grad limit ensures a constant mean free path and that the fraction of volume occupied by spheres is proportional to ε .

To study the Grad limit it is necessary to pass over to correlation functions. They are defined by

$$\rho_n^\varepsilon(x_1,\ldots,x_n,t) = \sum_{m=o}^{\infty} \frac{1}{m!} \int dx_{n+1} \ldots dx_{n+m}\, f_{n+m}^\varepsilon(x_1,\ldots,x_{n+m},t) \ . \qquad (3.1)$$

Here $f_n^\varepsilon(t)$ are the densities of the time-evolved measure $\mu^\varepsilon \circ T_{-t}^\varepsilon$. To ensure the existence of the Grad limit we impose two conditions on the correlation functions of the initial measure μ^ε . Let h_β be the normalized Maxwellian at inverse temperature β , $h_\beta(p) = (\beta/2\pi)^{3/2}[\exp-\beta p^2/2]$.

(C1) There exist constant M,z,β such that for all $n \geq 1$

$$\varepsilon^{2n}\rho_n^\varepsilon(x_1,\ldots,x_n) \leq M \prod_{j=1}^{n} \{zh_\beta(p_j)\} \quad . \tag{3.2}$$

(C2) There exist measurable functions $r_n : (\Lambda \times \mathbb{R}^3)^n \to \mathbb{R}$ such that for all $n \geq 1$

$$\lim_{\varepsilon\to 0} \varepsilon^{2n}\rho_n^\varepsilon(x_1,\ldots,x_n) = r_n(x_1,\ldots,x_n) \tag{3.3}$$

uniformly on compact sets of $(\Lambda \times \mathbb{R}^3)^n \setminus \{q_i = q_j \ , \ i,j = 1,\ldots,n\}$.

The bound (C1) ensures that μ^ε is uniquely determined by its correlation functions.

Lanford [2] has proved the following

Theorem 1. Let μ^ε be a sequence of probability measures on Γ such that its correlation functions satisfy (C1) and (C2). Then for $0 \leq t \leq 0.1\sqrt{\beta}/z$ and all $n \geq 1$

$$\lim_{\varepsilon\to 0} \varepsilon^{2n}\rho_n^\varepsilon(x_1,\ldots,x_n,t) = r_n(x_1,\ldots,x_n,t) \tag{3.4}$$

exist a.s. . The limit functions $r_n(t)$ are measurable and bounded as

$$0 \leq r_n(x_1,\ldots,x_n,t) \leq M' \prod_{j=1}^{n} \{z'h_{\beta'}(p_j)\} \tag{3.5}$$

with some (time-dependent) constants M',z',β' . They are the mild solution of the Boltzmann hierarchy

$$\frac{\partial}{\partial t} r_n(x_1,\ldots,x_{n,t}) = - \sum_{j=1}^{n} p_j \frac{\partial}{\partial q_j} r_n(x_1,\ldots,x_n,t)$$

$$+ \sum_{j=1}^{n} d^2 \int_{\hat\omega\cdot(p_{n+1})\geq 0} dp_{n+1} \hat\omega\cdot (p_{n+1}-p_j)$$

$$\{r_{n+1}(x_1,\ldots,q_j,p_j',\ldots,q_j,p_{n+1}',t) \\ -r_{n+1}(x_1,\ldots,q_j,p_j,\ldots,q_j,p_{n+1},t)\} \quad , \tag{3.6}$$

$n = 1,2,\ldots$.

Here (p_j',p_{n+1}') are a pair of incoming momenta with outgoing momenta (p_j,p_{n+1}) and $\hat{\omega}$ is the unit vector pointing from the j-th sphere to the (n+1)-th sphere.

Remarks. (i) Let $U_n(t)$ be the group induced by the flow of n independent particles on the torus Λ . Let the integral operator in (3.6) be denoted by $C_{j,n}$. Then by a mild solution we mean a solution to the integral equations

$$r_n(t) = U_n(t)r_n + \sum_{j=1}^{n} \int_0^t ds\, U_n(t-s)C_{j,n+1}\, r_{n+1}(s) \ , \tag{3.7}$$

$n = 1,2,\ldots$. Existence and uniqueness of solutions of (3.7) locally in time are established by iteration. If the initial data satisfy the bound (C1), then the solution satisfies the same bound but with time-dependent coefficients M',z',β' . These are guaranteed to be finite only for $0 \le t \le 0.1\sqrt{\beta}/z$.

(ii) Usually one assumes r_n to be continuous. In fact, measurability suffices and this turns out to be more convenient for our later purposes.

(iii) One has more precise information about the convergence (3.4). This is explained in the thesis of King [6].

(iv) Theorem 1 holds for a general class of boundary conditions, including stochastic boundary conditions as diffusive reflection.

(v) Related work is found in [7,8,9]. Time correlation functions at low density are discussed in [10,11,12].

What is the structure of the solution of the Boltzmann hierarchy ? If we choose arbitrary initial data subject only to symmetry and the bound (3.5), nothing beyond existence for short times can be said. However, one should remember that the initial data r_n are obtained through a particular limit procedure. Therefore we have to investigate the class of those r_n which, subject to conditions (C1) and (C2), can be reached as $\varepsilon \to 0$. For this class of initial data we will then show that the solution of the Boltzmann hierarchy can be constructed from the solution of the Boltzmann equation.

To determine the class of admissible r_n's looks like a hard analytical problem. Within the probabilistic framework, using the theory of random fields, a simple solution is obtainable.

Let $\mathcal{S}(\Lambda \times \mathbb{R}^3)$ be the Schwartz space of rapidly decreasing functions let $\mathcal{S}'(\Lambda \times \mathbb{R}^3)$ be the space of tempered distributions. $\mathcal{S}'(\Lambda \times \mathbb{R}^3)$ is equipped with its weak*-topology. For $f \in \mathcal{S}(\Lambda \times \mathbb{R}^3)$ and $n \in \mathcal{S}'(\Lambda \times \mathbb{R}^3)$ we denote by $n(f)$ the linear functional n evaluated at f. The linear functions on $\mathcal{S}'(\Lambda \times \mathbb{R}^3)$

$$n \mapsto n(f) \in \mathbb{R} \tag{3.8}$$

are continuous by definition.

Each point $(x_1,\ldots,x_n) \in \Gamma$ can be thought of as a particle configuration in $\Lambda \times \mathbb{R}^3$. To this configuration we associate an element of $\mathcal{S}'(\Lambda \times \mathbb{R}^3)$ by

$$S : (x_1,\ldots,x_n) \mapsto \varepsilon^2 \sum_{j=1}^{n} \delta_{x_j} \; . \tag{3.9}$$

δ_{x_j} is the Dirac delta-distribution (point measure) at $x_j \in \Lambda \times \mathbb{R}^3$. μ^ε gives a probability distribution of configurations of particles and therefore induces under S a probability measure $\langle\cdot\rangle_\varepsilon$ on $\mathcal{S}'(\Lambda \times \mathbb{R}^3)$ by

$$\langle\cdot\rangle_\varepsilon = \mu^\varepsilon \circ S^{-1} \tag{3.10}$$

with S^{-1} considered as a map for sets.

[A fine introduction to the mathematics of generalized random fields (equivalently to probability measures on $\mathcal{S}'(\Lambda \times \mathbb{R}^3)$) is the article by

M. Reed [13]. For our purpose we need only the fact that a probability measure $<\cdot>$ on $S'(\Lambda \times \mathbb{R}^3)$ is uniquely determined by its moments $<n(f_1)\ldots n(f_n)>$ provided they satisfy certain growth conditions. In the present case these growth conditions are always satisfied because of (C1).]

For n in the range of S we have

$$n(f) = \varepsilon^2 \sum_{j=1}^{n} f(x_j) \quad . \tag{3.11}$$

If f is replaced by the characteristic function of a set $\Delta \subset \Lambda \times \mathbb{R}^3$, then (3.11) is the (scaled) number of particles in Δ. The moments $<n(f_1)\ldots n(f_n)>_\varepsilon$ give the correlations between the total number of particles. The correlation functions as defined by (3.1) give correlations between n-tuples of different particles. Therefore $<n(f_1)\ldots n(f_n)>_\varepsilon$ depends linearly on the first n correlation functions. The first moment of $<\cdot>_\varepsilon$ is

$$<n(f)>_\varepsilon = \varepsilon^2 \int dx_1 \rho_1^\varepsilon(x_1) f(x_1) \quad . \tag{3.12}$$

The second moment is

$$\begin{aligned} <n(f_1)n(f_2)>_\varepsilon &= <\varepsilon^4 \sum_{i\neq j=1}^{n} f_1(x_i)f_2(x_j)>_\varepsilon \\ &+ <\varepsilon^4 \sum_{j=1}^{n} f_1(x_j)f_2(x_j)>_\varepsilon \\ &= \varepsilon^4 \int dx_1 dx_2 \rho_2^\varepsilon(x_1,x_2) f_1(x_1) f_2(x_2) \\ &+ \varepsilon^4 \int dx_1 \rho_1^\varepsilon(x_1) f_1(x_1) f_2(x_1) \quad . \end{aligned} \tag{3.13}$$

Similarly, the higher moments are

$$< \prod_{j=1}^{n} n(f_j)>_\varepsilon = \varepsilon^{2n} \int dx_1 \ldots dx_n \, \rho_n^\varepsilon(x_1,\ldots,x_n) \prod_{j=1}^{n} \{f_j(x_j)\} + O(\varepsilon^2) \quad . \tag{3.14}$$

These moments determine $<\cdot>_\varepsilon$ uniquely. (C1) and (C2) imply that as $\varepsilon \to 0$ the moments of $<\cdot>_\varepsilon$ converge and determine a unique limit measure $<\cdot>$. The moments of the limit measure are given by

$$\langle \prod_{j=1}^{n} n(f_j) \rangle = \int dx_1 \ldots dx_n \; r_n(x_1,\ldots,x_n) \prod_{j=1}^{n} f_j(x_j)$$
$$n = 1,2,\ldots, f_1,\ldots,f_n \in S(\Lambda \times \mathbb{R}^3) \; , \tag{3.15}$$

with the bound

$$0 \leq r_n(x_1,\ldots,x_n) \leq M \prod_{j=1}^{n} \{zh_\beta(p_j)\} \quad . \tag{3.16}$$

Theorem 2. A probability measure $\langle\cdot\rangle$ on $S'(\Lambda \times \mathbb{R}^3)$ which satisfies (3.15) and (3.16) is supported on the set $\Omega = \{r:\Lambda \times \mathbb{R}^3 \to \mathbb{R} \mid r$ is measurable and $0 \leq r \leq zh_\beta\} \subset S'(\Lambda \times \mathbb{R}^3)$.

Proof : Let $f \geq 0$. Then by (3.15) and (3.16)

$$0 \leq \langle n(f)^m \rangle \leq M[\int dqdp \; zh_\beta(p)f(q,p)]^m \tag{3.17}$$

for $m = 0,1,\ldots$. Therefore with probability one $0 \leq n(f) \leq \int dqdp \; z \, h_\beta(p)f(q,p)$. Since $S(\Lambda \times \mathbb{R}^3)$ has a countable basis, namely the Hermite functions, we conclude that $\langle\cdot\rangle$ - a.s.

$$0 \leq n(f) \leq \int dx_1 zh_\beta(p_1)f(x_1) \tag{3.18}$$

for all $f \in S(\Lambda \times \mathbb{R}^3)$ with $f \geq 0$. By (3.18) the map $f \mapsto n(f)$ extends to a positive, bounded linear functional on $L^1(\Lambda \times \mathbb{R}^3)$. Therefore there exists a function $r \in L^\infty(\Lambda \times \mathbb{R}^3)$ such that

$$n(f) = \int dx_1 r(x_1)f(x_1) \tag{3.19}$$

$\langle\cdot\rangle$ - a.s. and such that

$$0 \leq r(x_1) \leq zh_\beta(x_1) \quad \blacksquare \tag{3.20}$$

Let ν be the restriction of $\langle\cdot\rangle$ to Ω . Then Theorem 2 implies that

$$\langle n(f_1)\ldots n(f_n)\rangle = \int dx_1\ldots dx_n r_n(x_1,\ldots,x_n) \prod_{j=1}^{n} f_j(x_j)$$
$$= \int_\Omega \nu(dr) \prod_{j=1}^{n} \{ \int dx_j r(x_j)\ f_j(x_j)\} \quad . \tag{3.21}$$

In a more transparent form (3.21) is written as

$$r_n(x_1,\ldots,x_n) = \int_\Omega \nu(dr) \prod_{j=1}^{n} r(x_j) \quad . \tag{3.22}$$

Therefore r_n is just the n-th moment of the probability measure ν .

4. Boltzmann Hierarchy as Moment Equations.

The integral representation (3.22) allows us to interpret the Boltzmann hierarchy as the moment equations for the Boltzmann equation. Let $r(t) = T_t r$, $0 \le t \le 0.1\sqrt{\beta}/z$, be the mild solution of the Boltzmann equation

$$\frac{\partial}{\partial t} r(q,p,t) = - p \frac{\partial}{\partial q} r(q,p,t)$$
$$+ d^2 \int_{\hat\omega\cdot(p_1-p)\ge 0} dp_1\, d\hat\omega\, \hat\omega\cdot(p_1-p)\{r(q,q',t)r(q,p_1',t)-r(q,p,t)r(q,p_1,t)\} \tag{4.1}$$

with initial data $r \in \Omega$. It is known that then $0 \le T_t r \le z'h_{\beta'}$ with z' and β' depending on time and with z' and β' finite as long as time is restricted to $0 \le t \le 0.1\sqrt{\beta}/z$. We regard $T_t r \in S'(\Lambda \times \mathbb{R}^3)$. Then the semiflow T_t , $0 \le t \le 0.1\sqrt{\beta}/z$, is $\langle\cdot\rangle$ - a.s. defined.

Let us consider the moments

$$\langle \prod_{j=1}^{n} \{(T_t r)(f_j)\}\rangle = \int \nu(dr) \prod_{j=1}^{n} \int dx_j (T_t r)(x_j) f_j(x_j)$$
$$\equiv \int dx_1\ldots dx_n\, \tilde r(x_1,\ldots,x_n,t) \prod_{j=1}^{n} f_j(x_j) \quad .$$

We integrate the Boltzmann hierarchy in the form (3.7) over test functions and insert

(4.2). Using that $T_t r$ is the solution of the integral version of the Boltzmann equation one finds that $\tilde{r}(x_1,\ldots,x_n,t)$ is a solution of the Boltzmann hierarchy. By uniqueness of the solution $r_n(t) = \tilde{r}_n(t)$. Therefore the solution of the Boltzmann hierarchy is given by

$$r_n(x_1,\ldots,x_n,t) = \int_\Omega \nu(dr) \prod_{j=1}^{n} \{(T_t r)(x_j)\} \quad . \tag{4.3}$$

ν is the statistical distribution of the initial data for the Boltzmann equation. $r_n(x_1,\ldots,x_n,t)$ are the n-th moments of the statistical distribution at time t . Therefore solutions of the Boltzmann hierarchy are precisely the statistical solutions of the Boltzmann equation. Initial chaos means that the probability measure $\nu(dr)$ is concentrated on a single datum $r \in \Omega$. Propagation of chaos simply means that in this case the time evolved measure $\nu_t(dr)$ is concentrated on $T_t r$.

5. Speculations.

Are there physical situations in which the Boltzmann hierarchy rather than the Boltzmann equation has to be used ? According to our preceeding analysis this problem can be more sharply rephrased as : Are there physical situations where one is forced to use a statistical solution of the Boltzmann equation ? I do not know of any example. By analogy with simpler systems the following scenario has at least some degree of plausibility to me.

(i) In laboratory experiments the state prepared initially is usually a state of constrained thermal equilibrium. In particular, it is a state of local equilibrium. For these it is known that in the Grad limit initial chaos is satisfied [7] .

(ii) Although Theorem 1 is proved only under the restriction to short times, I believe that it is true for all times. Of course, the technical requirements might have to be modified. From the structure of the equations I therefore also must believe that, if initial chaos holds, then it is valid for all subsequent times. Nothing about uniformity in t is, and better should not be, claimed.

(ii) in conjunction with (i) seems to leave no room for statistical solutions of the Boltzmann equation. This is however not quite so.

(iii) In the qualitative theory of dissipative dynamical systems one uses invariant measures to describe the long time behavior of the system, cf. e.g. [14]. These measures are obtained through time averaging, weak - $\lim_{t\to\infty} \frac{1}{t} \int_0^t ds\; T_s x$. At least in principle such an analysis is also conceivable for the Boltzmann equation. If we consider a finite box with reflecting boundary conditions, then the solution of the Boltzmann equation should approach a global Maxwellian. In this case the time invariant measure is concentrated on that Maxwellian. However, if the system is driven by external sources, as e.g. thermal boundary conditions, in such a way that it does not approach a single stable stationary solution as $t \to \infty$, then non-degenerate invariant measures will exist. These measures define then non-factorized stationary solutions of the Boltzmann hierarchy with appropriate boundary conditions.

(iv) To be specific let us consider the Bernard problem (fluid layer heated from below, cooled from above and under the influence of gravity) in the turbulent regime at high Reynolds number. Let us model physical situation microscopically by a mechanical system of many particles subject to appropriate thermal boundary conditions. Under certain conditions on the intermolecular force the system approaches a unique steady state as $t\to\infty$ [15]. This state is, of course described by some probability measure μ^ε on Γ . It is conceivable that the correlation functions of μ^ε do not factorize in the Grad limit $\varepsilon \to 0$. In such a case the steady state correlation functions would tend at low density to a non-factorized stationary solution of the Boltzmann hierarchy with appropriate boundary conditions.

To some extent the situation here seems to be analogous to the one in fluid dynamics, cf. the investigation of stationary measures in [16], in particular the discussion at the end, and [17].

Acknowledgement. I am grateful to H. Grad for his insisting on the problem discussed here [18].

REFERENCES

[1] C. Cercignani, Transport Theory and Statistical Physics 2, 211 (1972).

[2] O.E. Lanford, Time Evolution of Large Classical Systems. In, Springer Lecture Notes in Physics, Vol. 38, p.1, ed. J. Moser, Springer, Berlin, 1975.

[3] H. Grad, private communication

[4] K. Alexander, Ph.D. Thesis, Department of Mathematics, University of California at Berkeley, 1975.

[5] M. Aizenman, Duke Math. J. 45, 809 (1978).

[6] F. King, Ph.D. Thesis, Department of Mathematics, University of California at Berkeley, 1975.

[7] O.E. Lanford, Soc. Math. de France, Astérisque 40, 117 (1976).

[8] O.E. Lanford, Lectures at the 3ième cycle, Ecole Polytechnique, Lausanne, unpublished.

[9] O.E. Lanford, Physica 106A, 70 (1981).

[10] H. van Beijeren, O.E. Lanford, J.L. Lebowitz, H. Spohn, J. Stat. Phys. 22, 237 (1980).

[11] H. Spohn, J. Stat. Phys., 26, 285 (1981)

[12] H. Spohn, Fluctuation Theory for the Boltzmann Equation, "Nonequilibrium Phenomena I. The Boltzmann Equation". Amsterdam, 1983.

[13] M. Reed, Functional Analysis and Probability Theory. In : Constructive Quantum Field Theory, ed. G. Velo and A. Wightman, Springer Lecture Notes in Physics, Vol. 25, p. . Springer, Berlin, 1973.

[14] O.E. Lanford, Qualitative and Statistical Theory of Dissipative Systems, Proceedings of the 1976 CIME school of statistical mechanics, Bressanone.

[15] S. Goldstein, J.L. Lebowitz, E. Presutti, Mechanical Systems with Stochastic Boundaries, Proceedings of Colloquium on "Random Fields : Rigorous Results in Statistical Mechanics and Quantum Field Theory", Esztergom, Hungary, 1979.

[16] D. Ruelle, Progress in Theor. Physics, Supplement 64, 339 (1978).

[17] D. Ruelle, Physics Letters 72A, 81 (1979).

[18] H. Grad, Singular Limits of Solutions of Boltzmann's Equation, Proceedings of the 8th International Symposium on Rarefied Gas Dynamics, Stanford, July 1972.

A NONLINEAR HALF-SPACE PROBLEM IN THE KINETIC THEORY OF GASES

Tor Ytrehus
The Norwegian Institute of Technology
7034 Trondheim - NTH, Norway

Abstract

The problem of the steady emission of a monatomic gas from a plane boundary into a half-space under finite pressure is considered. The use of a trimodal approximating distribution function and moments derived from the nonlinear Boltzmann equation leads to a four-moment system that can be solved exactly. The solution is naturally arrived at in two steps: First, an algebraic connection problem between the nonequilibrium gas state at the boundary and the external equilibrium state is solved. Then, the actual relaxation from the one state to the other is computed. For Maxwell molecules this latter problem has a particularly simple solution, but only for flow conditions such that the macroscopic gas velocity away from the boundary remains subsonic.

I. Introduction

In a number of actual physico-chemical gas flow situations one has to consider boundaries that emit and absorb gas molecules. Since the distribution function for the molecules arriving at such boundaries may be rather different from the one for molecules leaving, the state of the gas in the immediate vicinity of the boundaries may deviate significantly from a Maxwellian equilibrium. It is well known that the classical Hilbert and Chapman-Enskog expansions fail to give convergent solutions to the Boltzmann equation close to boundaries [1,2], and we are forced, therefore, to consider less rigorous methods when solving the governing kinetic equations in such regions of the flow.

The thickness of these kinetic boundary layers, or Knudsen layers, is of the order of a molecular mean free path and will, under ordinary conditions, be small compared to the linear extension of a boundary. Hence, the Knudsen layer problem may be treated one-dimensionally, and we are lead to study a half-space problem.

On the other hand, the absorption/emission Knudsen layers generally require nonlinear treatments due to large perturbation of the distribution function: Consider a plane boundary at x=0 emitting structureless molecules into the half-space x > 0 according to the distribution function

$$f_L^+ = \begin{cases} f_L \ , & \xi_x > 0 \\ 0 \ , & \xi_x < 0 \end{cases} \tag{1.1}$$

with f_L being a stationary wall Maxwellian

$$f_L = \frac{n_L}{(2\pi RT_L)^{3/2}} \exp\left(-\frac{\xi_x^2 + \xi_y^2 + \xi_z^2}{2RT_L}\right) \tag{1.2}$$

and ξ_x being the x-component of the molecular velocity. Let us assume that all molecules that are impinging upon the boundary from the half-space region are absorbed, and, furthermore, that their distribution function deviates from f_L by a perturbation $\phi_o f_L$. The net flux of molecules, $\dot{m}$, leaving the boundary can then be expressed as

$$\frac{\dot{m}}{n_L\sqrt{2RT_L}} = \int\limits_{\hat{\xi}_x < 0} \hat{\xi}_x \hat{f}_L \phi_o \, d\hat{\underline{\xi}} \tag{1.3}$$

where $\hat{\xi}_x$ and $\hat{f}_L \, d\hat{\underline{\xi}}$ have been normalized by the thermal velocity $\sqrt{2RT_L}$ and the number density n_L, respectively. The result (1.3) essentially shows that the nondimensional perturbation ϕ_o is of order of magnitude as the ratio of the net flux $\dot{m}$ of molecules out from the boundary, to the thermal flux $n_L\sqrt{2RT_L}$. In many typical absorption/emission problems of current interest, like strong evaporation and condensation at interphase boundaries, this ratio is by no means small: it may become of order unity in evaporation and even larger for condensation. The perturbation ϕ_o can, therefore, generally not be treated as a small quantity, and linear methods will become inadequate, at least sufficiently close to the interphase boundary.

Here we are going to describe a nonlinear solution to the strong evaporation half-space problem, based upon moment equations derived from the nonlinear Boltzmann equation for a monatomic gas. This may be considered an approximate solution to the Boltzmann equation itself. Essential to the method is the assumption of an approximation to the actual distribution function of the form

$$f(x,\vec{\xi}) = a_1^+(x) f_L^+(\vec{\xi}) + a_3^+(x) f_\infty^+(\vec{\xi}) + a_3^-(x) f_\infty^-(\vec{\xi}) \tag{1.4}$$

where f_L^+ is given in (1.1), (1.2) and $f_\infty^\pm$ are half-range Maxwellians corresponding to $\xi_x \gtrless 0$ in the equilibrium region outside of the Knudsen layer. The method combines features of the Mott-Smith [3] and Liu-Lees [4] treatments of shock structure and Couette flow problems, respectively, and it was developed independently in connection with work on laser induced metal evaporation [5,6], and in work on low density wind tunnel flow facilities [7,8]. Four moment equations are required to determine the amplitude functions $a_1^+(x)$, $a_\infty^{\pm}(x)$, and the unspecified elements (e.g., velocity and temperature) in the external drifting Maxwellian f_∞. It is a special feature of the method that it connects the states at the boundary and at downstream equilibrium by means of the conservation equations alone, somewhat similar to the Rankine-Hugoniot relations across a shock wave. This connection problem can be solved independent of the collision model and molecular interaction law for the gas. The Boltzmann H - theorem does, however, indicate [9] that the solution cannot be carried on to arbitrary large downstream flow velocities. Furthermore, the single non-conserved moment equation required to obtain the structure of the flow, and thus to complete the solution of the problem, in the case of Maxwell molecules and for the non-conserved quantity ξ_x^2, shows that the limiting downstream flow condition is very nearly sonic ($M_\infty = 0.993$ [9,10]). A solution to the nonlinear evaporation half-space problem, therefore, appears to exist at subsonic flow conditions, only. This conjecture is substantiated by numerical findings [11,12,13,14], and more explicitly by recent exact analytical results derived from the Boltzmann [15] - and Krook equation [16,17,18], after linearizing about the downstream drifting Maxwellian f_∞ (see also M. Arthur's contribution to this volume).

The present moment approach has been extended so as to include also the effects of internal degrees of freedom in polyatomic gases [19], and to the case of net condensation of a monatomic gas onto the interphase boundary [20].

II. Statement of Problem and Method of Solution

The situation referred to in the introductory part defines the following steady-state, half-space boundary value problem for the distribution function $f(x,\vec{\xi})$ in the nonlinear Boltzmann equation [1,2]

$$\xi_x \frac{\partial f}{\partial x} = \int (f'f_1' - ff_1)\, B(\theta,V)\, d\theta d\varepsilon\, \underline{d\xi_1} \tag{2.1}$$

$$f(0,\vec{\xi}) = f_L(\vec{\xi}) \ , \ \xi_x > 0, \tag{2.2}$$

where f_L is the wall Maxwellian (1.2), and $\vec{\xi}$ is a vector in three-dimensional velocity space, and where the notation otherwise is standard. The problem is defined for $x \geq 0$ and corresponds to a monatomic vapor gas being diffusely evaporated at an interphase boundary, x=0, according to the prescribed distribution function f_L, at the same time as all backscattered molecules are completely recondensed onto the boundary.

We assume that far downstream the gas relaxes to an equilibrium distribution characterized by a drift velocity u_∞, a number density n_∞ and a temperature T_∞; i.e.

$$\lim f(x,\vec{\xi}) = f_\infty(\vec{\xi}) = \frac{n_\infty}{(2\pi RT\ )^{3/2}} \exp\left(-\frac{(\xi_x-u_\infty)^2 + \xi_y^2+\xi_z^2}{2RT_\infty}\right) \qquad (2.3)$$

It is furthermore reasonable to assume that one of these parameters, say n_∞ (or $p_\infty = n_\infty kT_\infty$) is specified by conditions at "macroscopic infinity"; i.e. it is obtained in the solution of a macroscopic flow problem exterior to the kinetic boundary layer.

According to the introductory remarks, it is essential now to retain the nonlinearity in the problem, which is manifested in the products $f'f_1'$ and ff_1 in the collision term of the Boltzmann equation (2.1). Instead of attempting a direct solution of that equation, we are more or less forced to consider a simpler system, derived from the nonlinear Boltzmann equation, whose solution will represent a valid approximation to (2,1). Hence, we are led to the moment method [2,21], which in short consists in constructing an approximating distribution function

$$f(\vec{\xi},\ a_\nu)\ ,\ (\nu = 1,\ldots,\ M)$$

and then forming a sufficiently large number of moment equations from the Boltzmann equation to determine the unspecified parameters $a_\nu(\vec{x},t)$.

The moment equations are formed by multiplying both sides of the Boltzmann equation by functions $\psi_\mu(\vec{\xi})$ (μ=1,..., N) and integrating over velocity space; in the one-dimensional steady case (2.1) we obtain

$$\frac{\partial}{\partial x}\int \xi_x\psi_\mu(\vec{\xi})\ f\ \underline{d\xi} = \int \psi_\mu(\vec{\xi})Q(f,f_1)\underline{d\xi} \qquad (\mu=1,\ldots,N) \qquad (2.4)$$

where we have written $Q(f,f_1)$ for the nonlinear Boltzmann collision term in (2.1).

Making use of elementary properties of the collision operator [1,2,21], this system of moment equations (Maxwell-Boltzmann transfer equations) can be rewritten as

$$\frac{\partial}{\partial x}\int \xi_x \psi_\mu(\vec{\xi}) f \, \underline{d\xi} = \int (\psi_\mu' - \psi_\mu) f f_1 \, B(\theta, V) d\theta d\varepsilon \, \underline{d\xi} \, \underline{d\xi}_1 \qquad (2.5)$$

where $\psi_\mu' = \psi_\mu(\vec{\xi}')$ and $f_1 = f(\vec{\xi}_1)$. If we were to choose functions $\psi_\mu(\vec{\xi})$ ($\mu=1, \ldots, N$, $N \to \infty$) that formed a complete set, the system of infinitely many moment equations (2.5) would be equivalent to the Boltzmann equation (2.1) itself. However, from a practical point of view, we always have to consider only a finite number of transfer equations, and the hope is that, for sufficiently small N and a judicious choice of the arbitrary elements $a_\nu(\nu=1,\ldots,M)$ in the approximating distribution function, accurate results can still be obtained.

The actual choice is suggested by the physics in the problem at hand: for instance, the form of the distribution function (1.4) is chosen so as to exactly satisfy the boundary - and auxiliary conditions (2.2) and (2.3), and duly take into account the discontinuous behavior in velocity space (at $\xi_x = 0$). For this we impose the following conditions on the three amplitude functions $a_1^+(x)$, $a_3^+(x)$, $a_3^-(x)$ in (1.4)

$$x = 0: \quad \begin{aligned} a_1^+ &= 1 \\ a_3^+ &= 0 \\ a_3^- &= \beta^- \end{aligned} \qquad\qquad x \to \infty: \quad \begin{aligned} a_1^+ &= 0 \\ a_3^+ &= 1 \\ a_3^- &= 1 \end{aligned} \qquad (2.6)$$

where $\beta^- \equiv a_3^-(0)$ is an as yet unknown boundary value for the half-range mode f_∞^- at the surface. Thus, we are left with three unspecified x-dependent functions $a_1^+(x)$, $a_3^+(x)$, $a_3^-(x)$, and three unspecified constant parameters β^-, T_∞, u_∞ to be determined from the moment equations. It appears that the trimodal Ansatz (1.4) represents a minimum complexity required to adequately describe the present flow problem. Note that all the x-dependent parameters occur linearly in f, since the temperatures and the drift velocity in the Maxwellian basic modes f_L^+, $f_\infty^{\pm}$ are constants. This is a feature of the Mott-Smith method [3], which in its simplest version employs two basic full-range Maxwellian modes to describe the structure of a shock wave. The half-range character of the basic modes defined in (1.1) and (2.7) overleaf

$$f_\infty^+ = \begin{cases} f_\infty \ , \ \xi_x > 0 \\ 0 \ , \ \xi_x < 0 \end{cases} \quad , \quad f_\infty^- = \begin{cases} 0 \ , \ \xi_x > 0 \\ f_\infty \ , \ \xi_x < 0 \end{cases} \tag{2.7}$$

reflects a feature of the Liu-Lees method [4] devised to account for discontinuous behavior of the distribution function in the vicinity of solid boundaries. In this way the discontinuity is located exactly where predicted by free molecule theory ($\xi_x = 0$).

As the subsequent discussion will show, four moment equations are required to determine the three x-dependent parameters $a_\nu(x)$ in this case. Hence, N(=4) > M(=3), and some degeneracy must occur: Choosing for the first three generating functions ψ_μ the collisional invariants; i.e.

$$\psi_\mu = 1, \ \xi_x, \ \frac{1}{2}\xi^2 \ (\mu = 1, 2, 3), \tag{2.8}$$

we have from (2.5)

$$\frac{\partial}{\partial x}\int \xi_x \psi_\mu(\vec{\xi}) \ f \ \underline{d\xi} = 0 \quad (\mu = 1, 2, 3) \quad , \tag{2.9}$$

since the collisional contributions then vanishes. Writing the Ansatz (1.4) for f in the obvious notation

$$f = \sum_{\nu=1}^{3} a_\nu(x) \ f_\nu(\vec{\xi}) \ , \tag{2.10}$$

equations (2.9) integrates into the following linear algebraic system for the a_ν's

$$\sum_{\nu=1}^{3} a_\nu(x) \ B_{\mu\nu} = C_\mu \quad (\mu = 1, 2, 3) \quad , \tag{2.11}$$

where the coefficients $B_{\mu\nu}$ are given by

$$B_{\mu\nu} = \int \xi_x \ \psi_\mu(\vec{\xi}) f_\nu \ \underline{d\xi} \tag{2.12}$$

and the C_μ's are constants. If the system (2.11) were to posess a solution for $a_\nu(x)$ (ν = 1, 2, 3), except at the boundary x = 0, then the distribution function could be determined independent of collisional effects. Since this obviously cannot be the case, we must require instead

$$\det ||B_{\mu\nu}|| = 0, \tag{2.13}$$

and write one more moment equation; this time for a non-conserved quantity, for which we choose

$$\psi_4 = \xi_x^2 \qquad (\mu = 4) \tag{2.14}$$

Hence, we get a differential equation

$$\sum_{\nu=1}^{3} \frac{da_\nu}{dx} \cdot B_{4\nu} = \int (\xi_x'^2 - \xi_x^2) ff_1 \, B(\theta,V) d\theta d\varepsilon \, \underline{d\xi}_1 \, \underline{d\xi} \tag{2.15}$$

which by virtue of the product ff_1 in the collision term, is a bilinear equation in the a_ν's. The x-dependence of the parameters a_1^+, a_3^+, a_3^- is, therefore, determined from the four equations (2.11), (2.15), with the condition (2.13) taken into account. A simple approach to the solution is facilitated through a closer examination of the conservation equations (2.11) for the problem.

III. Gasdynamic Connection Problem

Writing out the conservation equations (2.11) in explicit terms, evaluating the constants C_μ at downstream equilibrium, we have

$$\begin{aligned} B_{11}a_1^+(x) + B_{12}a_3^+(x) + B_{13}a_3^-(x) &= n_\infty u_\infty \\ B_{21}a_1^+(x) + B_{22}a_3^+(x) + B_{23}a_3^-(x) &= n_\infty u_\infty^2 + n_\infty RT_\infty \\ B_{31}a_1^+(x) + B_{32}a_3^+(x) + B_{33}a_3^-(x) &= n_\infty u_\infty \left(\frac{1}{2} u_\infty^2 + \frac{5}{2} RT_\infty\right) \end{aligned} \tag{3.1}$$

where the coefficients $B_{\mu\nu}$ are given from (2.12) with the half-range Maxwellians f_L^+, f_∞^+, f_∞^- inserted for f_ν (ν = 1, 2, 3) as follows:

$$\begin{aligned} &B_{11} = n_L u_L \,, \quad B_{12} = n_\infty \sqrt{\frac{RT_\infty}{2\pi}}\, F^+, \quad B_{13} = -n_\infty \sqrt{\frac{RT_\infty}{2\pi}}\, F^- \\ &B_{21} = \frac{1}{2} n_L RT_L, \quad B_{22} = \frac{1}{2} n_\infty RT_\infty G^+ \,, \quad B_{23} = \frac{1}{2} n_\infty RT_\infty G^- \\ &B_{31} = 2n_L u_L RT_L, \quad B_{23} = 2n_\infty \sqrt{\frac{RT_\infty}{2\pi}}\, RT_\infty H^+ \,, \quad B_{33} = -2n_\infty \sqrt{\frac{RT_\infty}{2\pi}}\, RT_\infty H^- \end{aligned} \tag{3.2}$$

Here

$$n_L = \sqrt{\frac{RT_L}{2\pi}} \tag{3.3}$$

and the parameters $F^{\pm}$, $G^{\pm}$, $H^{\pm}$ have the definitions

$$F^{\pm} = \sqrt{\pi}\, S_\infty (\pm 1 + \text{erf } S_\infty) + e^{-S_\infty^2}$$

$$G^{\pm} = (2S_\infty^2 + 1)(1 \pm \text{erf } S_\infty) \pm \frac{2}{\sqrt{\pi}} S_\infty e^{-S_\infty^2} \tag{3.4}$$

$$H^{\pm} = \frac{\sqrt{\pi}\, S_\infty}{2} (S_\infty^2 + \frac{5}{2})(\pm 1 + \text{erf } S_\infty) + \frac{1}{2}(S_\infty^2 + 2)e^{-S_\infty^2}$$

in terms of the downstream speed ratio

$$S_\infty = \frac{u_\infty}{\sqrt{2RT_\infty}} \tag{3.5}$$

In the expressions (3.4) erf denotes the error function, i.e.:

$$\text{erf } S_\infty = \frac{2}{\sqrt{\pi}} \int_0^{S_\infty} e^{-t^2}\, dt \tag{3.6}$$

Also, the following relations among the plus - and minus functions are seen to apply

$$F^+ - F^- = 2\sqrt{\pi}\, S_\infty$$

$$G^+ + G^- = 4\, S_\infty^2 + 2$$

$$H^+ - H^- = \sqrt{\pi}\, S_\infty (S_\infty^2 + \frac{5}{2}) \tag{3.7}$$

If we write the equations (3.1) for x=0, i.e., at the boundary, using the boundary conditions in (2.6) on the amplitude functions a_1^+, a_3^+, a_3^-, we obtain, after suitably normalizing the resulting expressions

$$z_L \sqrt{\frac{T_\infty}{T_L}} - \beta^- F^- = 2\sqrt{\pi}\, S_\infty$$

$$z_L + \beta^- G^- = 4S_\infty^2 + 2$$

$$z_L - \beta^- \sqrt{\frac{T_\infty}{T_L}} H^- = \sqrt{\frac{T_\infty}{T_L}} \sqrt{\pi}\, S_\infty (S_\infty^2 + \frac{5}{2}) \tag{3.8}$$

Here $z_L (\geq 1)$ is a known driving parameter defined as the ratio between the pressure p_L associated with the wall Maxwellian f_L and the downstream equilibrium pressure p_∞; i.e.

$$z_L = \frac{p_L}{p_\infty} = \frac{n_L T_\infty}{n_\infty T_\infty} \tag{3.9}$$

Physically, p_L is the vapor saturation pressure corresponding to the dense phase temperature T_L of the boundary.

Since the system (3.8) relates the nonequilibrium vapor state at the boundary to the downstream equilibrium state, it constitutes a gasdynamic connection problem through the Knudsen layer, somewhat analogous to the Rankine-Hugoniot relations across a shock wave. A basic difference lies, of course, in the fact that the upstream state in the present case is at nonequilibrium and cannot be described at the Euler level of ordinary gas dynamics. Nevertheless, this state is, in the present description, independent of any collision model and molecular interaction law, since its description is fully contained in the conservation equations- with the assumption of a distribution function of the form (1.4).

A solution of the nonlinear system (3.8) yields the three parameters β^-, S_∞, $\sqrt{T_\infty/T_L}$ (T_L is given) as functions of the driving parameter z_L. Hence, the distribution function at the wall

$$f(0,\vec{\xi}) = \begin{cases} f_L , & \xi_x > 0 \\ \beta^- f_\infty , & \xi_x < 0 \end{cases} \tag{3.10}$$

is completely specified at this point. Therefore, the vapor nonequilibrium state n(0), T(0), u(0) is determined independent of the collisional properties of the gas. A particularly simple way to the solution of (3.8) was pointed out in [19]: Multiplying the first equation by $-\sqrt{T_\infty/T_L}$, the second by $\sqrt{\pi}\sqrt{T_\infty/T_L}\cdot S_\infty/4$, and adding to the third one, we obtain after dividing by z_L:

$$\frac{T_\infty}{T_L} + \frac{\sqrt{\pi}\, S_\infty}{4}\sqrt{\frac{T_\infty}{T_L}} - 1 = 0 \tag{3.11}$$

Discarding the negative root, we then have the result

$$\sqrt{\frac{T_\infty}{T_L}} = -\frac{\sqrt{\pi}\, S_\infty}{8} + \sqrt{1 + \frac{\pi}{64} S_\infty^2} \tag{3.12}$$

With this result the first two of equations (3.8) immediately yield

$$z_L = \frac{2\,e^{-S_\infty^2}}{F^- + \sqrt{\frac{T_\infty}{T_L}}\,G^-}\ , \quad \beta^- = \frac{(2S_\infty^2 + 1)\sqrt{\frac{T_\infty}{T_L}} - 2\sqrt{\pi}\,S_\infty}{F^- + \sqrt{\frac{T_\infty}{T_L}}\,G^-} \tag{3.13}$$

Although z_L is the independent variable here, the solution is most conveniently presented in the form above in terms of the speed ratio S_∞. The actual z_L-versus S_∞ relationship must be computed numerically anyway, and so must β^- versus S_∞. The state of the vapor gas at the boundary follows from the standard definitions

$$n = \int f\,\underline{d\xi}\ , \quad nu = \int \xi_x\,f\,\underline{d\xi}$$

$$\frac{3}{2}\,nkT = \int \frac{1}{2}\,m\,(\vec{\xi} - \vec{u})^2\,f\,\underline{d\xi}\ , \tag{3.14}$$

with the particular distribution function (3.10) inserted. For the density and the temperature at x=0 we obtain

$$\frac{n(0)}{n_L} = \frac{1}{2}\,[1 + \frac{n_\infty}{n_L}\,\beta^-(1 - \mathrm{erf}\,S_\infty)] \tag{3.15}$$

and

$$\frac{T(0)}{T_L}\quad \frac{1}{3}\,\frac{n_\infty T_\infty}{n(0)T_L}\,[2S_\infty^2\,(1 - \frac{n_\infty}{n(0)}) + 1 + z_L + \beta^-(1-\mathrm{erf}\,S_\infty)] \tag{3.16}$$

where the ratio n_∞/n_L is given by $z_L = P_L/P_\infty$ and T_∞/T_L; i.e.

$$\frac{n_\infty}{n_L} = \frac{1}{z_L}\,\frac{T_L}{T_\infty} \tag{3.17}$$

A summary of the results from the gasdynamic connection problem is contained in Table I overleaf, where also the fraction of backscattered molecules

$$\gamma = \int_{\xi_x < 0} |\xi_x|\,f(0,\vec{\xi})\,\underline{d\xi} \Bigg/ \int_{\xi_x > 0} \xi_x\,f(0,\vec{\xi})\,\underline{d\xi} \tag{3.18}$$

has been included.

Table I Summary of Gasdynamic Calculations

S_∞	z_L	T_∞/T_L	β^-	$T(0)/T_L$	γ
0	1.0000	1.0000	1.0000	1.0000	1.0000
0.1	1.2307	0.9567	1.0198	0.9607	0.7055
0.2	1.5002	0.9152	1.0597	0.9251	0.5060
0.3	1.8120	0.8756	1.1349	0.8950	0.3728
0.4	2.1698	0.8378	1.2711	0.8703	0.2860
0.5	2.5770	0.8016	1.5115	0.8509	0.2318
0.6	3.0369	0.7671	1.9284	0.8356	0.2004
0.7	3.5527	0.7342	2.6437	0.8230	0.1849
0.8	4.1273	0.7028	3.8625	0.8111	0.1804
0.9	4.7635	0.6729	5.9316	0.7989	0.1835
1.0	5.4639	0.6443	9.4405	0.7852	0.1917

The S_∞ - versus z_L relationships and the true kinetic temperature jump $T(0)/T_L$ are shown in comparison with accurate numerical solutions of the nonlinear Boltzmann and Krook equations [13], and with experimental results [9,10], in Figs. 1 and 2 below and overleaf.

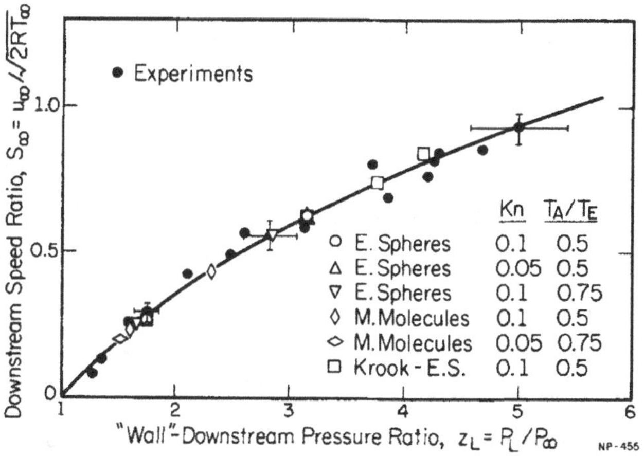

Fig. 1. Downstream speed ratio S_∞ versus pressure ratio $z_L = p_L/p_\infty$. (Open symbols refer to numerical Boltzmann - and Krook results for 1D-flow between absorbing/emitting parallel plates [13]. Closed symbols refer to effusion experiments [9,10].)

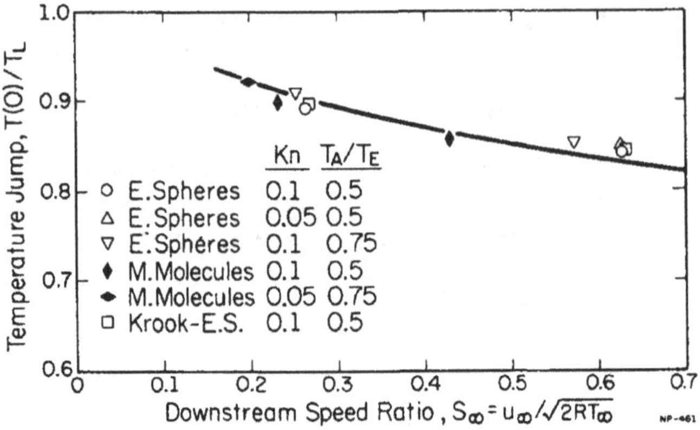

Fig. 2. Temperature jump $T(0)/T_L$ at evaporating surface versus downstream speed ratio S_∞. (Symbols refer to numerical Boltzmann - and Krook results for 1D-flow between absorbing/emitting parallel plates [13]).

It appears that the present treatment of the gasdynamic connection problem through the Knudsen layer as being independent of collision model and molecular interaction law, is essentially correct.

There is nothing in the above solution that prevents us from considering arbitrary large downstream speed ratios, as z_L evidently is a monotonic function of S_∞ by the first of equations (3.13). In particular, the flow is predicted to become supersonic ($S_\infty > \sqrt{5/6} \simeq 0.913$) above the critical value $(z_L)_{cr} \simeq 4.85$ of the pressure parameter. Since it would be unexpected to obtain highly supersonic equilibrium flow in a steady one-dimensional system, we may check the gasdynamic solution against the Boltzmann H-theorem [1,2] to see if it respects a general property of the collision term, which has been omitted so far in the calculations. The H-theorem for a bounded system under steady state conditions reads [1,2]

$$\int_S dS \int_{\vec{\xi}} (\vec{\nu} \cdot \vec{\xi})\, f \ln f \, \underline{d\xi} \leq 0 \qquad (3.19)$$

where the integral must be evaluated at the boundary S, having local outward directed normal $\vec{\nu}$. In the present case this leads to the inequality

$$\left(\int_{\vec{\xi}} \xi_x f \ln f \, \underline{d\xi}\right)_{x=0} - \left(\int \xi_x f \ln f \, \underline{d\xi}\right)_{x \to \infty} \geq 0 \tag{3.20}$$

Carrying out the integrations, using expressions (3.10), (1.2), (2.3) for the distribution function at the interphase boundary and at downstream equilibrium, we have, after normalizing with the number flux $n_L\sqrt{RT_L/2\pi}$

$$\ln \frac{n_L}{n_\infty}\left(\frac{T_\infty}{T_L}\right)^{3/2} + \beta^- \frac{n_\infty}{n_L}\sqrt{\frac{T_\infty}{T_L}}\,[F^- \ln\beta^- + 2e^{-S_\infty^2} - \frac{3\sqrt{\pi}}{2} S_\infty(1-\operatorname{erf} S_\infty)]$$

$$+ 3\frac{n_\infty}{n_L}\sqrt{\frac{T_\infty}{T_L}}\sqrt{\pi}\, S_\infty - 2 \geq 0 \tag{3.21}$$

The left-hand side of this equality, denoted as $\Delta h_x \equiv (h_x)_0 - (h_x)_\infty$, is plotted in Fig. 3 below, and it is seen to be positive for values of S_∞ that are below 1.20, only.

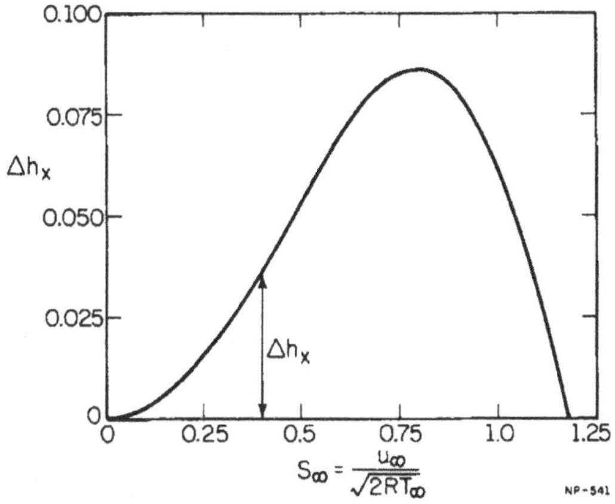

Figure 3. Results from Boltzmann H-theorem (H-theorem satisfied for $\Delta h_x \geq 0$, only).

The solution of the gasdynamic connection problem, therefore, has no physical significance at high supersonic downstream flow velocities. In accordance with the general view that the Boltzmann H-theorem poses a weak condition on the parameters in a nonequilibrium flow problem, we shall find in the subsequent discussion of the detailed structure of the flow in the Knudsen layer, that the upper limit in speed ratio is even more restrictive, such that the solution to our problem is meaningful at subsonic flow conditions, only. Nevertheless, the present case is an example among very few, in which the H-theorem gives a little more than purely qualitative information.

IV. Knudsen Layer Structure

Coming back to the problem of determining the x-dependent elements in the distribution function as posed in Section II, we need to work out the details in the nonconserved moment equation (2.15). Before doing so, we recall an important property of the system of conserved moment equations (3.1): Because of the condition (2.13), the rank of the matrix of that system is 2, and one arbitrary function $a_\nu(x)$ will appear in the solution. This means that, for instance, $a_1^+(x)$ and $a_3^+(x)$ may be expressed in terms of $a_3^-(x)$. Simple manipulations in the system, using the relations (3.7) lead to the following results

$$a_1^+(x) = \frac{a_3^-(x) - 1}{\beta^- - 1} \quad , \quad a_3^+(x) = \frac{\beta^- - a_3^-(x)}{\beta^- - 1} \quad , \tag{4.1}$$

which also imply the simple relation

$$a_1^+(x) + a_3^+(x) = 1 \tag{4.2}$$

It is readily concluded that the condition (2.13) for the above solution (4.1) is automatically satisfied by the relations (3.12), (3.13) from the gasdynamic connection problem (3.8). Hence, the Rankine-Hugoniot like equations in the gasdynamic problem constitute a necessary condition for the solution of the full kinetic problem to exist, in complete analogy with the Mott-Smith approach for the shock structure problem.

Let us first treat the convective terms in the nonconserved equation, noting that the coefficients $B_{4\nu}$ for $\psi_4 = \xi_x^2$ are determined as follows:

$$B_{41} = 2n_L u_L RT_L \qquad B_{42} = 2n_\infty \sqrt{\frac{RT_\infty}{2\pi}}\, RT_\infty\, K^+$$

$$B_{43} = 2n_\infty \sqrt{\frac{RT_\infty}{2\pi}} RT_\infty K^-, \tag{4.3}$$

in terms of functions $K^{\pm}(S_\infty)$ of the downstream speed ratio S_∞

$$K^{\pm} = (S_\infty^2 + \frac{3}{2})(\pm 1 + \text{erf } S_\infty) + (1 + S_\infty^2)\, e^{-S_\infty^2} \tag{4.4}$$

Now, using the relations (4.1) and the obvious property

$$K^+ - K^- = \sqrt{\pi}\, S_\infty(2S_\infty^2 + 3) \tag{4.5}$$

the convective term is expressed as

$$\sum_{\nu=1}^{3} \frac{da_\nu}{dx} B_{4\nu} = \frac{1}{\beta^- - 1} \{2n_L u_L RT_L - 2n_\infty \sqrt{\frac{RT_\infty}{2\pi}} RT_\infty [\sqrt{\pi}\, S_\infty(2S_\infty^2 + 3) + \beta^- K^-]\} \frac{da_3^-}{dx} \tag{4.6}$$

It is possible (although not essential) to greatly simplify this result by observing the additional property

$$K^- = 2H^- - F^- \quad , \tag{4.7}$$

and then using the last two of the Rankine-Hugoniot like equations (3.8) to eliminate H^- and F^-. Important cancelling then occurs, and the final result for the convective term is

$$\sum_{\nu=1} \frac{da_\nu}{dx} B_{4\nu} = - \frac{2n_L u_L RT_L}{\beta^- - 1} (1 - \frac{T_\infty}{T_L}) \frac{da_3^-}{dx} \tag{4.8}$$

Then, proceeding to the collision term in (2.15) we note that, with the Ansatz (1.4), the term can be worked out as a bilinear form in $a_\nu a_\ell$ for any molecular interaction law for which the indicated integrals in velocity space exist. A general simplification occurs for inverse power repulsive forces ($F = kr^{-n}$) for which we have [1,2]

$$B(\theta,V) = V^{n-5/n-1}\, \Theta\,(\theta) \tag{4.9}$$

i.e. the dependence of B upon θ and V factorizes. Choosing furthermore n=5, the dependence upon the relative velocity disappears altogether, and we have the particular case of <u>Maxwell molecules</u>. For this interaction model the integrals of the collision term can be expressed in terms of moments up to the order $\int \psi_4 f \, d\underline{\xi}$, for any ψ_4 beyond the collision invariants. For instance, we have [22]

$$\int (\xi_x'^2 - \xi_x^2) f f_1 \;\; \Theta \;\; (\theta) d\theta d\varepsilon \; \underline{d\xi_1} \underline{d\xi} = \frac{\pi}{\lambda_{rm}} \sqrt{\frac{RT_r}{2\pi}} \; \frac{n}{n_r} \; \tau'_{xx} \tag{4.10}$$

where τ'_{xx} denotes the viscous part of the normal stress in the x-direction:

$$\tau'_{xx} = - m[\int (\xi_x - u)^2 f \; \underline{d\xi} - \frac{1}{3} \int (\vec{\xi}-\vec{u})^2 \; f \; \underline{d\xi} \;] \tag{4.11}$$

and subscript r refers to a convenient reference state. The mean free path λ_r is related to the Chapman-Enskog viscosity μ_r by

$$\lambda_r = \frac{\mu_r}{mn_r} \sqrt{\frac{\pi}{2RT_r}} \quad , \tag{4.12}$$

m being the molecular mass. Making use of the Ansatz (1.4) we now write

$$\tau'_{xx} = \sum_{\nu=1}^{3} a_\nu (\tau'_{xx})_\nu \tag{4.13}$$

where a modal contribution is defined by

$$(\tau'_{xx})_\nu = m \; [\int c_x c_x f_\nu \; \underline{dc} - \frac{1}{3} \int c^2 f_\nu \underline{dc} \;] \quad , \tag{4.14}$$

and where we have introduced the thermal velocity

$$c_x = \xi_x - u(x) \tag{4.15}$$

Taking into account symmetry around the c_x axis in velocity space, we furthermore have

$$(\tau'_{xx})_\nu = - \frac{2}{3} m[\int c_x c_x f_\nu \; \underline{dc} - \int c^2 \; f_\nu \; \underline{dc} \;] \quad , \tag{4.16}$$

and the contributions from the different half-range modes are evaluated with the following results

$$(\tau'_{xx})_L = - \frac{2}{3} m[\frac{1}{2} n_L(-4uu_L + u^2)]$$

$$(\tau'_{xx})^{\pm}_{\infty} = - \frac{2}{3} m[\frac{1}{2} n_\infty RT_\infty (1 \mp \frac{2S_\infty}{\sqrt{\pi}} e^{-S_\infty^2} \pm \text{erf } S_\infty)$$

$$\pm \; 2(u_\infty - u) n_\infty \sqrt{\frac{RT_\infty}{2\pi}} \; e^{-S_\infty^2}$$

$$+ \; (u_\infty - u)^2 \frac{1}{2} n_\infty (1 \pm \text{erf } S_\infty) - \frac{1}{2} n_\infty RT_\infty (1 \pm \text{erf } S_\infty)] \tag{4.17}$$

Here u is the local x-dependent macroscopic velocity according to definitions (3.14):

$$nu = n_\infty u_\infty \quad , \quad n = \int f \, \underline{d\xi} = \sum_{\nu=1}^{3} a_\nu(x) \int f_\nu \, \underline{d\xi} \tag{4.18}$$

The total viscous stress is then computed from (4.13), and we have, for the product $n\tau'_{xx}$ occuring in the collision term (4.10):

$$n\tau'_{xx} = - \frac{\frac{2}{3} m(n_\infty u_\infty)^2}{4S_\infty^2} \left[-4S_\infty^2 + \frac{n}{n_\infty} \{a_3^+[G^+ - (1 + \text{erf } S_\infty)] + a_3^-[G^- - (1 - \text{erf } S_\infty)]\}\right] \tag{4.19}$$

Then, using the momentum conservation equation and the relations (4.1) we finally arrive at the bilinear form in the amplitude function $a_3^-(x)$:

$$n\tau'_{xx} = \frac{\frac{2}{3} m(n_\infty u_\infty)^2}{8S_\infty^2} \phi_1\phi_2[a_3^-(x) - 1][a_3^-(x) - r] \tag{4.20}$$

with $\phi_1\phi_2$ and r being defined as

$$\phi_1 = \frac{1}{\beta^- - 1} \left[\frac{n_L}{n_\infty} - 2 + \beta^-(1 - \text{erf } S_\infty)\right]$$

$$\phi_2 = \frac{1}{\beta^- - 1} \left[z_L - 2 + \beta^-(1 - \text{erf } S_\infty)\right]$$

$$r = 1 - \frac{2}{\phi_1} + \frac{4S_\infty^2}{\phi_2} \tag{4.21}$$

The moment equation (2.15) for the non-conserved quantity ξ_x^2 and for Maxwell molecules thus, by virtue of the results (4.8), (4.10) and (4.20), finally reads

$$\frac{da_3^-}{dx} = - \frac{A}{\lambda_L} (a_3^- - 1)(a_3^- - r)$$

$$A = \frac{\pi}{12} \left(\frac{n_\infty}{n_L}\right)^2 \frac{T_\infty/T_L}{1-T_\infty/T_L} \tag{4.22}$$

with λ_L being the molecular mean free path (4.12) corresponding to the Maxwellian state n_L, T_L. Since the remaining two amplitude functions a_1^+ and a_3^+ immediately follow from (4.1) once a_3^- has been determined, the complete structure of the Knudsen layer is contained in the differential equation (4.22) above.

The equation has two critical points: $a_3^- = 1$, corresponding to downstream equilibrium, and $a_3^- = r$. Now, the range of the function $a_3^-(x)$ is clearly $[1,\beta^-]$ (note, Table I, that $\beta^- \geq 1$), and the equation (4.22), therefore, implies a relaxation towards downstream equilibrium only if the parameter r is outside of that range. A plot of the parameter r versus the downstream speed ratio S_∞ is given in Fig. 4 below,

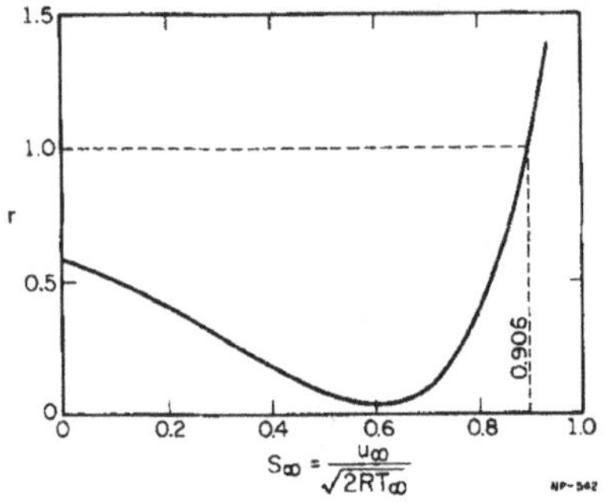

Fig. 4. Critical point in ξ_x^2 - moment equation (solution exists for $r \leq 1$, only).

and it is seen that r is indeed outside the range, for values of S_∞ on the interval $[0,(S_\infty)_{cr.}]$, only, where $(S_\infty)_{cr.}$ is approximately equal to 0.906. Hence, there is a critical Mach number $(M_\infty)_{cr.}$

$$(M_\infty)_{cr.} = \frac{6}{5}(S_\infty)_{cr.} \approx 0.993 \tag{4.23}$$

above which there is no solution to the evaporation Knudsen layer problem. At flow conditions below that Mach number, the Knudsen layer structure is simply given by the integral of (4.22) subject to the boundary condition $a_3^-(0) = \beta^-$

$$\frac{a_3^-(x)-1}{a_3^-(x)-r} = \left(\frac{\beta^- -1}{\beta^- -r}\right) \exp\left\{-A\frac{(1-r)}{\lambda_L} x\right\} \tag{4.24}$$

Macroscopic variables then follow from standard definitions; for instance, we have by (3.14) for density and temperature

$$\frac{n}{n_\infty} = [a_3^-(x)-1]\,\frac{1}{2}\,\phi_1 + 1$$

$$\frac{T}{T_\infty} = \frac{1}{3}\frac{n_\infty}{n}[2S_\infty^2 + 3 + [a_3^-(x)-1]\phi_2 - 2S_\infty^2\frac{n_\infty}{n}] \tag{4.25}$$

with ϕ_1 and ϕ_2 being defined in (4.21).

A proper scale for the thickness of the Knudsen layer, therefore, appears to be

$$\ell = \frac{\lambda_L}{A(1-r)} \quad , \tag{4.26}$$

in which the quantity A(1-r) remains of order unity, except near $S_\infty \simeq (S_\infty)_{cr}$. There the thickness blows up as $r \to 1$ and the downstream equilibrium is not reached within a distance from the boundary comparable to the mean free path λ_L. This behavior at critical flow conditions is qualitatively confirmed by the Monte Carlo results [12], and it is furthermore reminiscent of the behavior of shock-wave thickness for the Mach number approaching unity.

The critical value (0.993) for the Mach number M_∞ is sufficiently close to unity for the conjecture to be made that the problem (2.1), (2.2), (2.3) has a unique solution if and only if $M_\infty \leq 1$. This conjecture, along with a first step towards a proof based upon the Boltzmann equation linearized about f_∞, was put forward in [15]. Subsequently, the conjecture has been proven for a strictly one dimensional BGK model [16,17], and very recently for the three dimensional BGK model [18], also by linearizing the distribution function about the downstream Maxwellian f_∞.

To exhibit the large perturbation and the discontinuous behavior in the distribution function that generally occur across an absorption/emission kinetic boundary layer, the reduced distribution function

$$\Phi(x,\xi_x) = \int f\, d\xi_y\, d\xi_z \;/\; n_L(2\pi RT_L)^{-1/2}$$

$$= \begin{cases} a_1^+(x)e^{-\xi_x^2/2RT_L} + \dfrac{n_\infty}{n_L}\left(\dfrac{T_L}{T_\infty}\right)^{1/2} a_3^+(x)e^{-(\xi_x-u_\infty)^2/2RT_\infty} & , \ \xi_x > 0 \\ \\ \dfrac{n_\infty}{n_L}\left(\dfrac{T_L}{T_\infty}\right)^{1/2} a_3^-(x)\, e^{-(\xi_x-u_\infty)^2/2RT_\infty} & , \ \xi_x < 0 \end{cases} \tag{4.27}$$

is shown plotted overleaf for a typical case of $S_\infty = u_\infty/\sqrt{2RT_\infty} = 0.5$.

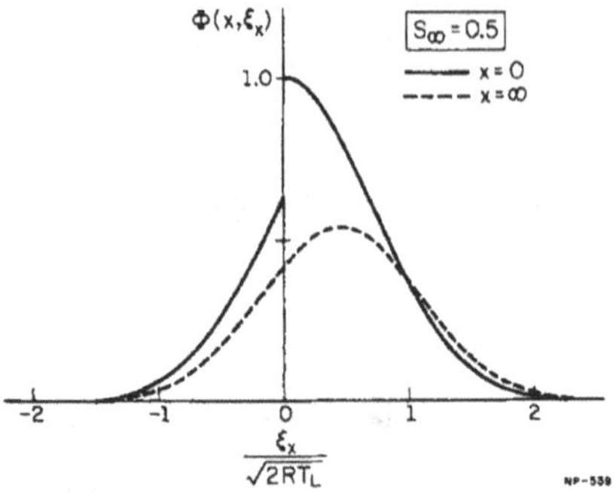

Fig. 5. Reduced distribution function $\Phi(x,\xi_x)$ at boundary and at downstream equilibrium for typical flow conditions $S_\infty = 0.5$.

It is to be expected that linearized treatments must become inadequate in dealing with the total transition across this type of Knudsen layer, because the perturbation in the distribution function is by no means small. Thus, the linear results in [17,18] show important discrepancies with the present results, and with accurate numerical results [13] for the gasdynamic connection problem. across the Knudsen layer, although they provide a valid description of the flow behavior close to downstream equilibrium.

Finally, it should be remarked that the Knudsen layer structure in the present moment method turns out to be very sensitive to the choice of the non-conserved quantity ψ_4, [23]: For $\psi_4 = \xi_x^3$ we get a non-conserved moment equation similar to (4.22), but with different parameters A and r, such that the second critical point approaches unity at flow conditions close to $S_\infty = 0.60$ ($M_\infty \simeq 0.65$). Below this value of the speed ratio the thickness scale (4.26) is increased compared to the ξ_x^2 - calculations, whereas above this value there is no solution to the Knudsen layer problem. For $\psi_4 = \xi_x \xi^2$ the non-conserved moment equation (surprisingly) becomes linear and the criticality condition thereby gets lost. The equation implies a much weaker approach towards downstream equilibrium than in previous cases, resulting in a much thicker Knudsen layer. An even more severe degeneracy occurs for this velocity moment also in the Mott-Smith method for the shock structure problem, as noted in [24]. It may appear from this that the Knudsen layer structure is more satisfactorily predicted by the lower velocity moments. A possible reason for this result could be that inaccuracies in the approximating distribution function, especially at the higher velocities, will be more heavily weighted and therefore amplified by the higher moments.

V. Concluding Remarks

The present moment solution of the nonlinear evaporation half-space problem shows that a large portion of the solution can be obtained from the conservation equations alone, once a judicious choice for the approximating distribution function has been made.

Some practical usefulness of the Boltzmann H-theorem for a semi-bounded system has been demonstrated in assessing the limiting flow conditions for the problem. Our findings agree with the general idea that the H-theorem is a necessary, but not sufficiently strong condition to ensure that useful solutions to the full set of moment equations exist.

The particular simplification of Maxwell molecules is manifested in a simple, exact solveable non-conserved transfer equation, which yields the structure of the Knudsen layer and contains a condition for the existence of a solution. Obviously, this part of the solution is not unique, since it depends upon the choice of non-conserved velocity moment. There are at least some reasons to believe that the lower degree moments yield the better approximation to the Boltzmann equation, for a given approximating distribution function. In particular, the moment equation for ξ_x^2 yields a criticality condition that is very close to the one obtained from the Boltzmann and BGK equations with exact methods.

References

1. C. Cercignani, "Mathematical Methods in Kinetic Theory," Plenum Press, New York, and McMillan, London (1969).

2. C. Cercignani, "Theory and Application of the Boltzmann Equation," Scottish Academic Press, Edinbourgh, and Elsevier, New York (1975).

3. H. M. Mott-Smith, Phys. Rev. 82, 892 (1951).

4. C. H. Liu and L. Lees, in "Rarefied Gas Dynamics," p. 391, L. Talbot, ed. Academic Press, New York-London (1961).

5. S. I. Anisimov, Soviet Physics JETP, 27, 182 (1969).

6. A. V. Luikov, T. L. Perleman and S. I. Anisimov, Int. J. Heat and Mass Transfer, 14, 177 (1971).

7. T. Ytrehus, J. J. Smolderen and J. Wendt, Entropie, 42, 33 (1971).

8. T. Ytrehus, in "Rarefied Gas Dynamics," B.4, M. Becker and M. Fiebig, eds. DVLFR Press, Portz-Wahn (1974).

9. T. Ytrehus, von Karman Institute Technical Note 112, Rhode-St. Genese (1975).

10. T. Ytrehus, in "Rarefied Gas Dynamics," p. 1197, J. L. Potter, ed., AIAA, New York (1977).

11. M. N. Kogan and N. K. Makashev, Fluid Dynamics, 6, 913 (1974).

12. M. Murakami and K. Oshima, in "Rarefied Gas Dynamics," F.6, M. Becker and M. Fiebig, eds., DFVLR Press, Portz-Wahn (1974).

13. S. M. Yen and T. J. Akai, in "Rarefied Gas Dynamics," p. 1175, J. L. Potter, ed., AIAA, New York (1977).

14. T. Tran Cong and G. A. Bird, Phys. Fluids, 21, 327 (1978).

15. C. Cercignani, in "Mathematical Methods in Kinetic Theory of Gases," p. 129, D. C. Pack and H. Neunzert, eds., Peter D. Lang, Frankfurt A. M., Bern, Cirencester/U.K. (1980).

16. M. D. Arthur and C. Cercignani, ZAMP, 31, 634 (1980).

17. C. E. Siewert and J. R. Thomas, Jr., ZAMP, 32, 421, (1981).

18. C. E. Siewert and J. R. Thomas, Jr., ZAMP, 33, 473, (1982).

19. C. Cercignani, in "Rarefied Gas Dynamics," p. 305, S. S. Fisher, ed., AIAA, New York (1981).

20. T. Ytrehus and J. Alvestad, in "Rarefied Gas Dynamics," p. 330, S. S. Fisher, ed., AIAA, New York (1981).

21. M. N. Kogan, "Rarefied Gas Dynamics," Plenum Press, New York (1969).

22. W. G. Vincenti and C. H. Kruger, Jr., "Introduction to Physical Gas Dynamics," p. 364, J. Wiley, New York-London (1965).

23. T. Ytrehus, "Theoretical and Experimental Study of a Kinetic Boundary Layer Produced by Effusion from a Perforated Wall," Doctoral Thesis, Vrije Universiteit Brussel, Brussels (1975).

24. L. H. Holway, Jr., in "Rarefied Gas Dynamics," p. 193, J. H. dee Leew, ed., Academic Press, New York-London (1965).